Questions marked with the symbol should be attempted without using a calculator.

Prime Factors, HCF and LCM

Prime Factors

Apart from **prime numbers**, any whole number greater than 1 can be written as a product of **prime factors.** This means the number is written using only prime numbers multiplied together.

A prime number has only two factors, 1 and itself. 1 is not a prime number.

The prime numbers up to 20 are:

2, 3, 5, 7, 11, 13, 17, 19

The diagram below shows the prime factors of 60.

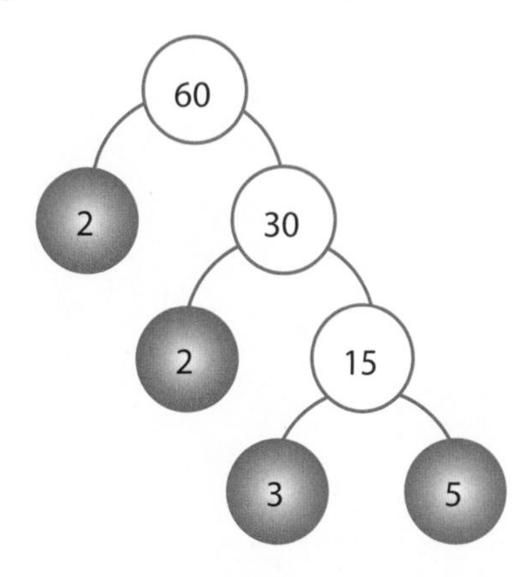

- Divide 60 by its first prime factor, 2.
- Divide 30 by its first prime factor, 2.
- Divide 15 by its first prime factor, 3.
- We can now stop because the number 5 is prime.

As a product of its prime factors, 60 may be written as:

$60 = 2 \times 2 \times 3 \times 5$

or in index form

$60 = 2^2 \times 3 \times 5$

Highest Common Factor (HCF)

The highest factor that two numbers have in common is called the **HCF**.

Example
Find the HCF of 60 and 96.

- Write the numbers as products of their prime factors.

$60 = 2 \times 2 \qquad \times 3 \times 5$

$96 = 2 \times 2 \times 2 \times 2 \times 2 \times 3$

- Ring the factors that are common.

$60 = \boxed{2} \times \boxed{2} \qquad \times \boxed{3} \times 5$

$96 = \boxed{2} \times \boxed{2} \times 2 \times 2 \times 2 \times \boxed{3}$

- These give the HCF $= 2 \times 2 \times 3$

$= \mathbf{12}$

GCSE In a Week

Maths

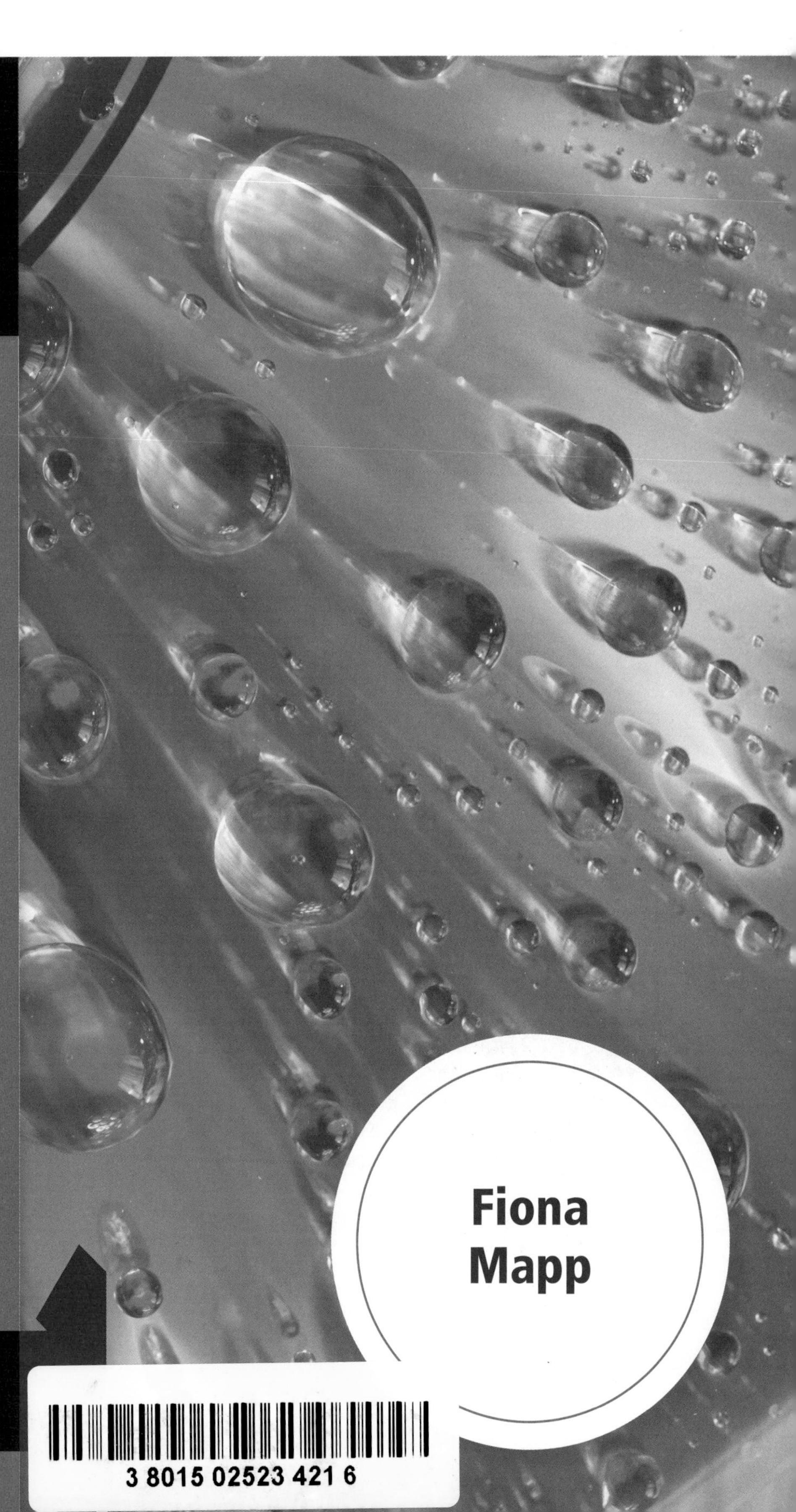

Fiona Mapp

Foundation Tier

Revision Planner

DAY 1

Page	Time	Topic	Date	Time	Completed
4	15 minutes	Prime Factors, HCF and LCM			
6	20 minutes	Fractions and Decimals			
8	15 minutes	Rounding and Estimating			
10	15 minutes	Indices			
12	20 minutes	Standard Index Form			
14	20 minutes	Formulae and Expressions 1			
16	20 minutes	Formulae and Expressions 2			

DAY 2

Page	Time	Topic	Date	Time	Completed
18	15 minutes	Brackets and Factorisation			
20	15 minutes	Equations 1			
22	20 minutes	Equations 2			
24	20 minutes	Simultaneous Linear Equations			
26	20 minutes	Sequences			
28	15 minutes	Inequalities			

DAY 3

Page	Time	Topic	Date	Time	Completed
30	20 minutes	Straight-line Graphs			
32	25 minutes	Curved Graphs			
34	20 minutes	Percentages			
36	20 minutes	Repeated Percentage Change			
38	20 minutes	Reverse Percentage Problems			

DAY 4

Page	Time	Topic	Date	Time	Completed
40	15 minutes	Ratio and Proportion			
42	20 minutes	Proportionality			
44	20 minutes	Measurement			
46	15 minutes	Interpreting Graphs			
48	15 minutes	Similarity			
50	15 minutes	2D and 3D Shapes			

Lowest (Least) Common Multiple (LCM)

The **LCM** is the lowest number that is a multiple of two numbers.

Example
Find the LCM of 60 and 96.

- Write the numbers as products of their prime factors.

$60 = 2 \times 2 \qquad\qquad \times 3 \times 5$

$96 = 2 \times 2 \times 2 \times 2 \times 2 \times 3$

- 60 and 96 have a common factor of $2 \times 2 \times 3$, so it is only counted once.

$60 = \boxed{2} \times \boxed{2} \qquad\qquad \times \boxed{3} \times 5$

$96 = \boxed{2} \times \boxed{2} \times 2 \times 2 \times 2 \times \boxed{3}$

- The LCM of 60 and 96 is

$2 \times 2 \times 2 \times 2 \times 2 \times 3 \times 5$

$=$ **480**

SUMMARY

- **Any whole number greater than 1 can be written as a product of its prime factors, apart from prime numbers themselves (1 is not prime).**
- **The highest factor that two numbers have in common is called the highest common factor (HCF).**
- **The lowest number that is a multiple of two numbers is called the lowest (least) common multiple (LCM).**

QUESTIONS

QUICK TEST

1. Write these numbers as products of their prime factors:

   a. 50 **b.** 360 **c.** 16

2. Decide whether these statements are true or false:

   a. The HCF of 20 and 40 is 4.

   b. The LCM of 6 and 8 is 24.

   c. The HCF of 84 and 360 is 12.

   d. The LCM of 24 and 60 is 180.

EXAM PRACTICE

1. Find the highest common factor of 120 and 42. [3 marks]

2. Buses to St Albans leave the bus station every 20 minutes. Buses to Hatfield leave the bus station every 14 minutes.

   A bus to St Albans and a bus to Hatfield both leave the bus station at 10 am.

   When will buses to both St Albans and Hatfield next leave the bus station at the same time? [3 marks]

Fractions and Decimals

A **fraction** is part of a whole number. The top number is the **numerator** and the bottom number is the **denominator**. Fractions can be cancelled by dividing the numerator and denominator by a common factor:

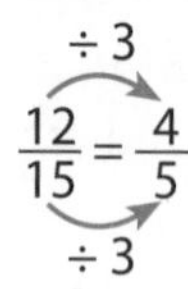

Addition

You need to change the fractions so that they have the same denominator.

Example

$\frac{5}{9} + \frac{1}{7}$

$= \frac{35}{63} + \frac{9}{63}$ — The lowest common denominator is 63 since both 9 and 7 go into 63.

$= \mathbf{\frac{44}{63}}$ — Remember to add only the numerators and not the denominators.

Subtraction

You need to change the fractions so that they have the same denominator.

Example

$\frac{4}{5} - \frac{1}{3}$ — The lowest common denominator is 15.

$= \frac{12}{15} - \frac{5}{15}$

$= \mathbf{\frac{7}{15}}$ — Remember to subtract only the numerators and not the denominators.

Multiplication

Before starting, write out whole or mixed numbers as improper fractions (also known as top-heavy fractions).

Example

$\frac{2}{7} \times \frac{4}{5}$

$= \frac{2 \times 4}{7 \times 5}$ — Multiply the numerators together. Multiply the denominators together.

$= \mathbf{\frac{8}{35}}$

Division

Before starting, write out whole or mixed numbers as improper fractions.

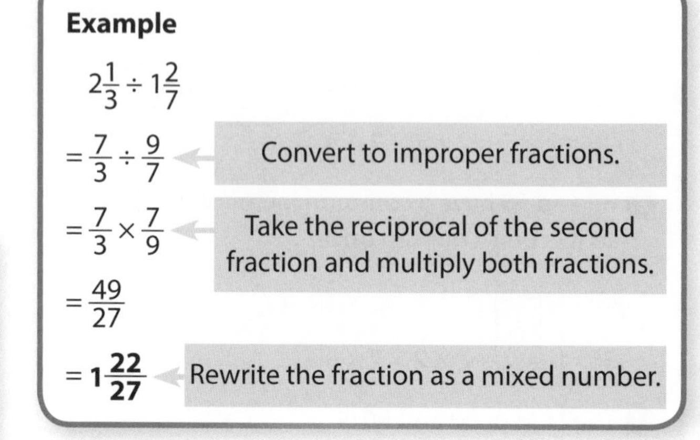

Reciprocals

The **reciprocal** of a number $\frac{x}{a}$ is $\frac{a}{x}$

For example, the reciprocal of $\frac{4}{7}$ is $\frac{7}{4}$

Decimals and Fractions

To change a fraction into a decimal, divide the numerator by the denominator, either by short division or by using a calculator.

To change a decimal into a fraction, write the decimal as a fraction with a denominator of 10, 100, etc. (look at the last decimal place to decide) and then cancel.

Examples

$\frac{2}{5} = 2 \div 5 = \mathbf{0.4}$

$\frac{1}{8} = 1 \div 8 = \mathbf{0.125}$

$0.23 = \mathbf{\frac{23}{100}}$ — The last decimal place is 'hundredths' so the denominator is 100.

$0.165 = \frac{165}{1000} = \mathbf{\frac{33}{200}}$

Decimals that never stop and have a repeating pattern are called **recurring** decimals. All fractions give either **terminating** or **recurring** decimals.

Examples

$\frac{1}{3} = 0.333\,3333\ldots$ usually written as $\mathbf{0.\dot{3}}$

$\frac{5}{11} = 0.454\,545\,45\ldots$ usually written as $\mathbf{0.\dot{4}\dot{5}}$

$\frac{4}{7} = 0.571\,428\,571\ldots$ usually written as $\mathbf{0.\dot{5}71\,42\dot{8}}$

Fraction Problems

You may need to solve problems involving fractions.

Examples

1. A school has 1400 pupils. 740 pupils are boys.

   $\frac{3}{5}$ of the boys and $\frac{1}{4}$ of the girls study French.

   Work out the total number of pupils in the school who study French.

   $\frac{3}{5} \times 740 = 444$ boys study French ← Work out the number of boys who study French.

   $1400 - 740 = 660$ are girls ← Work out the number of girls in the school.

   $\frac{1}{4} \times 660 = 165$ girls study French

   $444 + 165 =$ **609 pupils study French**

2. Charlotte's take-home pay is £930. She gives her mother $\frac{1}{3}$ of this and spends $\frac{1}{5}$ of the £930 on going out. What fraction of the £930 is left?

   Give your answer as a fraction in its simplest form.

   $\frac{1}{3} + \frac{1}{5}$ ← This is a simple addition of fractions question.

   $= \frac{5}{15} + \frac{3}{15}$ ← Write the fractions with a common denominator.

   $= \frac{8}{15}$

   $1 - \frac{8}{15}$ ← You need the fraction of the money that is left, so subtract $\frac{8}{15}$ from 1.

   $= \mathbf{\frac{7}{15}}$ ← The fraction is in its simplest form.

SUMMARY

- **To add or subtract fractions, write them using the same denominator.**
- **To multiply fractions, multiply the numerators and multiply the denominators.**
- **To divide fractions, take the reciprocal of the second fraction and multiply the fractions together.**
- **When multiplying and dividing fractions, write out whole or mixed numbers as improper fractions before you begin.**
- **Decimals that never stop and have a repeating pattern are recurring decimals.**

QUESTIONS

QUICK TEST

1. Work out the following:

   a. $\frac{2}{3} + \frac{1}{5}$ **b.** $2\frac{6}{7} - \frac{1}{3}$

   c. $\frac{2}{9} \times \frac{5}{7}$ **d.** $\frac{3}{11} \div \frac{22}{27}$

2. Work out the following:

   a. $2\frac{1}{2} + 3\frac{1}{5}$ **b.** $2\frac{7}{10} - 1\frac{1}{9}$

   c. $3\frac{1}{5} \times \frac{2}{15}$ **d.** $5\frac{1}{4} \div \frac{3}{8}$

EXAM PRACTICE

1. In a magazine $\frac{3}{7}$ of the pages have advertisements on them. Given that 12 pages have advertisements on them, work out the number of pages in the magazine. [2 marks]

2. Rosie watches two television programmes. The first programme is $\frac{3}{4}$ of an hour and the second is $2\frac{2}{3}$ hours long. Work out the total length of the two programmes. [3 marks]

3. Place these fractions and decimals in order of size, smallest first:

   $\frac{3}{4}$ 0.4 0.85 $\frac{8}{10}$ $\frac{675}{1000}$ [2 marks]

Rounding and Estimating

Rounding to the Nearest 10, 100 and 1000

When rounding you must always look at the digit in the next place value column. If the digit is 5 or more, round up. If the digit is 4 or less, you round down.

For example, 2469.1 is:

- 2469 to the nearest whole number
- 2470 to the nearest 10
- 2500 to the nearest 100
- 2000 to the nearest 1000.

Decimal Places

When rounding numbers to a given number of **decimal places** (d.p.), count the number of places to the right of the decimal point, then look at the next digit on the right.

If the number is 5 or bigger, round up.
If the number is 4 or smaller, the digit stays the same.

2.3725 = 2.373 (3 d.p.)

The digit is 5 so round up the 2.

The 2 rounds up to a 3.

Examples
Round the following to the number of decimal places specified in the brackets.

1. 4.6931 (2 d.p.) = **4.69** — The 3 is less than 5, so the 9 does not change.
2. 27.325 (2 d.p.) = **27.33** — The 5 has the effect of rounding the 2 to a 3.
3. 149.3867 (3 d.p.) = **149.387** — The 7 has the effect of rounding the 6 to a 7.
4. 271.74 (1 d.p.) = **271.7** — The 4 is less than 5, so has no effect on the 7.

Significant Figures

The first **significant figure** (s.f.) is the first digit that is not a zero. The 2nd, 3rd… significant figures follow on after the first significant figure. They may or may not be zeros.

The same rules apply as in decimal places.

6347 = 6350 (3 s.f.)

The digit is bigger than 5, so the 4 rounds to a 5.

You must fill in the end zero(s). This is often forgotten.

Examples

1. Round 9.3156 to…
   - a. 3 s.f. = **9.32** — The 5 has the effect of rounding the 1 to a 2.
   - b. 2 s.f. = **9.3** — The 1 is less than 5, so the 3 does not change.
   - c. 1 s.f. = **9** — The 3 is less than 5, so to 1 s.f. it is 9, since 9.3156 is nearer to 9 than 10.
2. Round 0.735 to…
   - a. 2 s.f. = **0.74** — The 5 has the effect of rounding the 3 to a 4.
   - b. 1 s.f. = **0.7** — The 3 is less than 5, so the 7 does not change.

Estimating

When estimating the answer to a calculation, you must round each number to 1 significant figure.

$$\frac{273 \times 49}{28} \approx \frac{300 \times 50}{30} = \mathbf{500}$$

≈ means approximately equal to

If a measurement is accurate to some given amount, then the true value lies within half a unit of that amount.

The **upper bound** is the maximum possible value the measurement could have been.

The **lower bound** is the minimum possible value the measurement could have been.

If the weight (w) of a cat is 8.3 kg to the nearest tenth of a kilogram, then the weight would lie between 8.25 kg and 8.35 kg.

The limits of accuracy can be written using inequalities as shown:

$$\mathbf{8.25} \leqslant w < \mathbf{8.35}$$

Lower bound (8.25) — Upper bound (8.35)

SUMMARY

- **To round or correct to a given number of decimal places (d.p.), count that number of decimal places to the right of the decimal point. Look at the next digit on the right. If it is 5 or more you need to round up. Otherwise the digit stays the same.**
- **When measurements are given to a certain degree of accuracy:**
  - **highest possible value = upper bound**
  - **lowest possible value = lower bound**
- **For any number, the first significant figure is the first number that is not a zero. The 2nd, 3rd... significant figures follow on after the first significant figure. They may or may not be zeros.**

QUESTIONS

QUICK TEST

1. Put a ring around the correct answer.
   3724 rounded to 2 significant figures is:

   3800 37 38 3700

2. Decide whether the following statements are true or false:

   a. 4625 rounded to 3 s.f. is 4630

   b. 2.795 rounded to 1 d.p. is 2.7

   c. 0.00527 rounded to 2 s.f. is 0.0053

   d. 37 062 has 4 significant figures

EXAM PRACTICE

1. Work out an estimate for:
   $$\frac{306 \times 2.93}{0.051}$$ [3 marks]

2. The weight of a book is 28 grams to the nearest gram. Write down the lower bound of the weight of the book. [1 mark]

Indices

An **index** is sometimes called a **power**.

The base → a^b ← The index or power

Laws of Indices

The laws of indices can be used for numbers or algebra. The base has to be the same when the laws of indices are applied.

$$a^n \times a^m = a^{n+m}$$

$$a^n \div a^m = a^{n-m}$$

$$(a^n)^m = a^{n \times m}$$

$$a^0 = 1$$

$$a^1 = a$$

$$a^{-n} = \frac{1}{a^n}$$

$$a^{\frac{1}{m}} = \sqrt[m]{a}$$

$$a^{\frac{n}{m}} = (\sqrt[m]{a})^n$$

Examples with Numbers

1. Simplify the following, leaving your answers in index notation.

a. $5^2 \times 5^3 = 5^{2+3} = \mathbf{5^5}$

b. $8^{-5} \times 8^{12} = 8^{-5+12} = \mathbf{8^7}$

c. $(2^3)^4 = 2^{3\times4} = \mathbf{2^{12}}$

2. Evaluate:

Evaluate means to work out.

a. $4^2 = 4 \times 4 = \mathbf{16}$

b. $5^0 = \mathbf{1}$

c. $3^{-2} = \frac{1}{3^2} = \mathbf{\frac{1}{9}}$

d. $36^{\frac{1}{2}} = \sqrt{36} = \mathbf{6}$

e. $8^{\frac{2}{3}} = (\sqrt[3]{8})^2 = 2^2 = \mathbf{4}$

3. Simplify the following, leaving your answers in index form.

a. $7^2 \times 7^5 = \mathbf{7^7}$

b. $6^9 \div 6^2 = \mathbf{6^7}$

c. $\frac{3^7 \times 3^2}{3^{10}} = \frac{3^9}{3^{10}} = \mathbf{3^{-1}}$

d. $7^9 \div 7^{-10} = \mathbf{7^{19}}$

4. Evaluate:

a. $3^3 = 3 \times 3 \times 3 = \mathbf{27}$

b. $7^0 = \mathbf{1}$

c. $64^{\frac{1}{3}} = \sqrt[3]{64} = \mathbf{4}$

d. $81^{\frac{1}{2}} = \sqrt{81} = \mathbf{9}$

e. $5^{-2} = \frac{1}{5^2} = \mathbf{\frac{1}{25}}$

f. $\left(\frac{4}{9}\right)^{-2} = \left(\frac{9}{4}\right)^2 = \frac{81}{16} = \mathbf{5\frac{1}{16}}$

Examples with Algebra

1. Simplify the following:

a. $a^4 \times a^{-6} = a^{4-6} = a^{-2} = \mathbf{\frac{1}{a^2}}$

b. $5y^2 \times 3y^6 = \mathbf{15y^8}$

The numbers are multiplied.

The indices are added.

c. $(4x^3)^2 = \mathbf{16x^6}$

Remember to square the 4 and multiply the indices.

If in doubt, write it out: $(4x^3)^2 = 4x^3 \times 4x^3$

$= \mathbf{16x^6}$

d. $(3x^4y^2)^3 = \mathbf{27x^{12}y^6}$

or $3x^4y^2 \times 3x^4y^2 \times 3x^4y^2 = \mathbf{27x^{12}y^6}$

e. $(2x)^{-3} = \frac{1}{(2x)^3} = \mathbf{\frac{1}{8x^3}}$

2. Simplify:

a. $\frac{15b^4 \times 3b^7}{5b^2} = \frac{45b^{11}}{5b^2} = \mathbf{9b^9}$

b. $\frac{16a^2b^4}{4ab^3} = \mathbf{4ab}$

3. Simplify:

a. $7a^2 \times 3a^2b = \mathbf{21a^4b}$

b. $\frac{14a^2b^4}{7ab} = \mathbf{2ab^3}$

c. $\frac{9x^2y \times 2xy^3}{6xy} = \frac{18x^3y^4}{6xy}$

$= \mathbf{3x^2y^3}$

SUMMARY

- **Make sure you know and can use all the laws of indices.**
- **A negative power is the reciprocal of the positive power.**
- **Fractional indices mean roots.**

QUESTIONS

QUICK TEST

1. Simplify the following, leaving your answers in index form.

a. $6^3 \times 6^5$

b. $12^{10} \div 12^{-3}$

c. $(5^2)^3$

d. $64^{\frac{2}{3}}$

2. Simplify the following:

a. $2b^4 \times 3b^6$

b. $8b^{-12} \div 4b^4$

c. $(3b^4)^2$

d. $(5x^2y^3)^{-2}$

EXAM PRACTICE

1. Evaluate:

a. 5^0 [1 mark]

b. 7^{-2} [1 mark]

c. $64^{\frac{1}{3}} \times 144^{\frac{1}{2}}$ [2 marks]

d. $27^{-\frac{2}{3}}$ [2 marks]

2. Simplify:

a. $\frac{x^4 \times x^7}{x^{15}}$ [2 marks]

b. $\frac{3x^4 \times 4x^2}{2x^3}$ [2 marks]

Standard Index Form

Standard index form (standard form) is useful for writing very large or very small numbers in a simpler way.

When written in standard form a number will be written as:

$$a \times 10^n$$

A number between 1 and 10: $1 \leq a < 10$

The value of n is the number of places the digits have to be moved to return the number to its original value.

If the number is 10 or more, n is positive.

If the number is less than 1, n is negative.

If the number is 1 or more but less than 10, n is zero.

Examples

1. Write 2 730 000 in standard form.
   - 2.73 is the number between 1 and 10 ($1 \leq 2.73 < 10$)
   - Count how many spaces the digits have to move to restore the original number.

     The digits have moved 6 places to the left because it has been multiplied by 10^6

     2 . 7 3

     2 7 3 0 0 0 0

     So, 2 730 000 = $\mathbf{2.73 \times 10^6}$

2. Write 0.000 046 in standard form.
   - Put the decimal point between the 4 and 6, so the number lies between 1 and 10.
   - Move the digits five places to the right to restore the original number.
   - The value of n is negative.

     So, 0.000 046 = $\mathbf{4.6 \times 10^{-5}}$

On a Calculator

To put a number written in standard form into your calculator, you use the following keys:

[$\times 10^x$] [EXP] or [EE]

For example, $(2 \times 10^3) \times (6 \times 10^7) = 1.2 \times 10^{11}$ would be keyed in as:

[2] [$\times 10^x$] [3] [$\times$] [6] [$\times 10^x$] [7] [=]

or [2] [EXP] [3] [$\times$] [6] [EXP] [7] [=]

Doing Calculations

Examples

Work out the following using a calculator. Check that you get the answers given here.

1. $(6.7 \times 10^7)^3 = \mathbf{3.0 \times 10^{23}}$ (2 s.f.)
2. $\dfrac{(4 \times 10^9)}{(3 \times 10^4)^2} = \mathbf{4.\dot{4}}$
3. $\dfrac{(5.2 \times 10^6) \times (3 \times 10^7)}{(4.2 \times 10^5)^2} = \mathbf{884.4}$ (1 d.p.)

Examples
On a non-calculator paper you can use indices to help work out your answers.

1. $(2 \times 10^3) \times (6 \times 10^7)$

$= (2 \times 6) \times (10^3 \times 10^7)$

$= 12 \times 10^{3+7}$

$= 12 \times 10^{10}$

$= 1.2 \times 10^1 \times 10^{10}$

$= \mathbf{1.2 \times 10^{11}}$

2. $(6 \times 10^4) \div (3 \times 10^{-2})$

$= (6 \div 3) \times (10^4 \div 10^{-2})$

$= 2 \times 10^{4-(-2)}$

$= \mathbf{2 \times 10^6}$

3. $(3 \times 10^4)^2$

$= (3 \times 10^4) \times (3 \times 10^4)$

$= (3 \times 3) \times (10^4 \times 10^4)$

$= \mathbf{9 \times 10^8}$

You also need to be able to work out more complex calculations.

Example
The mass of Saturn is 5.7×10^{26} tonnes. The mass of the Earth is 6.1×10^{21} tonnes. How many times heavier is Saturn than the Earth? Give your answer in standard form, correct to 2 significant figures.

$$\frac{5.7 \times 10^{26}}{6.1 \times 10^{21}} = 93\,442.6$$

Now rewrite your answer in standard form.

Saturn is 9.3×10^4 times heavier than the Earth.

SUMMARY

- **Numbers in standard form will be written as $a \times 10^n$.**
- **$1 \leq a < 10$**
- **n is positive when the original number is 10 or more.**
- **n is negative when the original number is less than 1.**
- **n is zero when the original number is 1 or more but less than 10.**

QUESTIONS

QUICK TEST

1. Write in standard form:
   a. 64 000
   b. 0.000 46
2. Work out the following calculations. Leave in standard form.
   a. $(3 \times 10^4) \times (4 \times 10^6)$
   b. $(6 \times 10^{-5}) \div (3 \times 10^{-4})$
3. Work these out on a calculator:
   a. $(4.6 \times 10^{12}) \div (3.2 \times 10^{-6})$
   b. $(7.4 \times 10^9)^2 + (4.1 \times 10^{11})$

EXAM PRACTICE

1. a. Write 40 000 000 in standard form. [1 mark]
   b. Write 6×10^{-5} as an ordinary number. [1 mark]
2. The mass of an atom is 2×10^{-23} grams. What is the total mass of 7×10^{16} of these atoms?

   Give your answer in standard form. [3 marks]

Formulae and Expressions 1

A **term** is a collection of numbers, letters and brackets, all multiplied together, e.g. $6a$, $2ab$, $3(x - 1)$.

Expressions are made up of a number of terms, e.g. $a + 6$.

Terms are separated by + and – signs. Each term has a + or – attached to the front of it.

$$5ab - 3c - 6b^2 + 7$$

Invisible + sign | ab term | c term | b^2 term | Number term

Term	What it means
$3c$	$3 \times c$ **or** $c \times 3$ **or** $c + c + c$
ab	$a \times b$ **or** $b \times a$
b^2	b multiplied by itself $= b \times b$
$3b^2$	$3 \times b \times b$

- $a \div 2$ can be written as $\frac{a}{2}$
- $c \times a \times 5 = 5ac$; the number usually comes first and then the letters in alphabetical order
- $3a^2$ is not the same as $(3a)^2$

  $3a^2$ is 3 lots of just a^2

  $(3a)^2$ is 3 multiplied by a, then all of it squared.

Collecting Like Terms

Expressions can be simplified by collecting **like terms**.

Simplify means make the expression simpler.

You can only collect together terms that include exactly the same letter combinations.

Examples

1. $5a + 3a = \mathbf{8a}$
2. $3a - 4b + 2a + 3b = \mathbf{5a - b}$

   Work out $3a + 2a$ first and then $-4b + 3b$, which is $-1b$ or $-b$.
3. $5a^2 + 3a^2 - 2a^2 = \mathbf{6a^2}$
4. $5a - 3b$ cannot be simplified.
5. $2ab + 3ba = \mathbf{5ab}$

   ab means the same as ba.
6. Write down an expression for the perimeter of the shape. Give your answer in its simplest form.

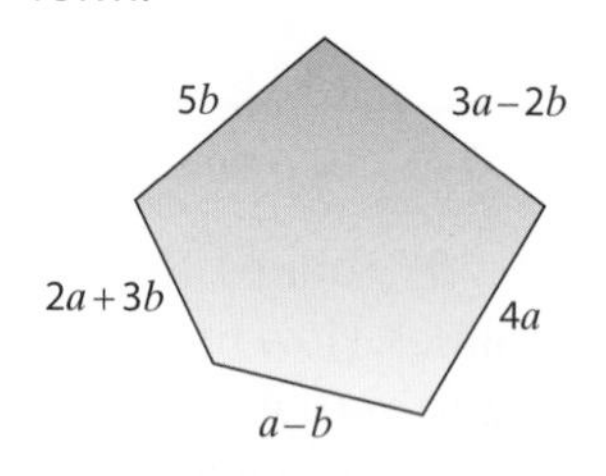

Adding all the lengths together gives:

$= 5b + 3a - 2b + 4a + a - b + 2a + 3b$

$= \mathbf{10a + 5b}$

Writing Formulae

$b = a + 6$ is an example of a **formula**. The value of b depends on the value of a.

Sometimes you will need to write your own formulae.

Example

My brother is three years older than me. My mother is three times as old as me.

a. If I am n years old, write expressions for my brother's and mother's ages.

If I am 12, my brother is $12 + 3 = 15$

Try using numbers first.

So if I am n years old, my brother is **$n + 3$ years old**.

If I am 12, my mother is $3 \times 12 = 36$

So if I am n years old, my mother is **$3 \times n$** or **$3n$ years old**.

b. Write down a formula for the sum (s) of the ages of me, my mother and brother.

$s = n + n + 3 + 3n$

$\mathbf{s = 5n + 3}$

SUMMARY

- **Expressions are made up of a number of terms added or subtracted together.**
- **Collect like terms to simplify an expression.**
- **You can only collect together terms that include exactly the same letter combinations.**

QUESTIONS

QUICK TEST

1. Simplify the following expressions:

   a. $a + a + a + a$

   b. $5a + 2b + 3a - b$

   c. $6a - 3b + 2a - 4b$

   d. $12xy + 4xy - xy$

   e. $3a^2 - 6b^2 - 2b^2 + a^2$

   f. $5xy - 3yx + 2xy^2$

EXAM PRACTICE

1. Simplify the following expressions:

   a. $5ab - 2bc + 6bc - 7ab$ [2 marks]

   b. $d \times d \times d \times d$ [1 mark]

   c. $5m \times 3n$ [1 mark]

2. Lauren buys x books costing £7 each and y magazines costing 98p each. Write down a formula for the total cost (T) of the books and magazines. [3 marks]

3. Adult cinema tickets cost £x and child cinema tickets cost £y. Mr Khan buys 2 adult tickets and 4 child tickets. Write down a formula in terms of x and y for the total cost (£C) of the tickets. [3 marks]

Formulae and Expressions 2

Substituting into Formulae

Replacing a letter with a number is called **substitution**. Write out the expression first and then replace the letters with the values given.

Work out the value but take care with the order of operations, i.e. **BIDMAS**.

Examples

1. $a = 3b - 4c$. Find a if $b = 4$ and $c = -2$

$a = (3 \times 4) - (4 \times -2)$
$= 12 - (-8)$ ← Taking away a negative is the same as adding.
$= \mathbf{20}$

2. $E = \frac{1}{2}mv^2$. Find E if $m = 6$ and $v = 10$

$E = \frac{1}{2} \times 6 \times 10^2$
$= \frac{1}{2} \times 6 \times 100$
$= \mathbf{300}$

3. $V = u + at$. Find V if $u = 22$, $a = -2$ and $t = 6$

$V = 22 + (-2 \times 6)$
$= 22 + (-12)$
$= 22 - 12$
$= \mathbf{10}$

4. $S = kt^2$. Find S when $k = 4$ and $t = -3$

$S = 4 \times (-3)^2$
$S = 4 \times 9$
$S = \mathbf{36}$

5. The cost of hiring a car can be worked out using this rule: Cost = £85 + 48p per mile.

a. Josh hires a car and drives 130 miles. Work out the cost.

C = £85 + 48p per mile
$= 85 + 0.48 \times 130$ ← Change 48p into £ so they are in the same unit: 48p = £0.48
$= 85 + 62.40$
= **£147.40**

b. On a different day Josh pays £131.56 for hiring the car. How many miles did he travel?

$131.56 = 85 + 0.48 \times$ number of miles
$131.56 - 85 = 46.56$
$\frac{46.56}{0.48} =$ **97 miles** ← Divide by the pence per mile to find the number of miles.

Rearranging Formulae

The subject of a formula is the letter that appears on its own on one side of the formula.

Any letter in a formula can become the subject by rearranging the formula.

Examples

1. Make c the subject of the formula:
$b = c - a$

$b = c - a$

$b + a = c$ ← Add a to both sides.

So $\boldsymbol{c = b + a}$

Examples (cont.)

2. Make a the subject of the formula:
$b = (a - 3)^2$

$b = (a - 3)^2$ — Deal with the power first. Square root both sides of the formula.

$\pm\sqrt{b} = a - 3$ — Remove any term added or subtracted. In this case add 3 to both sides of the formula.

$\pm\sqrt{b} + 3 = a$

$\boldsymbol{a = \pm\sqrt{b} + 3}$ — When square rooting you get a positive and negative solution, which is shown as $\pm$

3. Make x the subject of the formula:
$5(y + x) = 8x + 3$

When the subject occurs on both sides of the equals sign, they need to be collected on one side.

$5(y + x) = 8x + 3$

$5y + 5x = 8x + 3$

$5y - 3 = 8x - 5x$ — Collect the x terms on one side.

$5y - 3 = 3x$

$\boldsymbol{x = \frac{5y - 3}{3}}$ — Divide both sides by 3 to make x the subject.

4. Make x the subject of the formula $a = \frac{x + c}{x - d}$

$a = \frac{x + c}{x - d}$ — If the subject appears twice, you are likely to need to factorise once all the required terms are on one side.

$a(x - d) = x + c$ — Multiply both sides by $(x - d)$

$ax - ad = x + c$ — Multiply out the brackets.

$ax - x = c + ad$ — Collect like terms involving x on one side.

$x(a - 1) = c + ad$ — Factorise.

$\boldsymbol{x = \frac{c + ad}{a - 1}}$

SUMMARY

- **Substitution is replacing letters in a formula with numbers.**
- **Any letter can become the subject of a formula by rearranging the formula.**
- **When rearranging a formula, you must do the same thing to both sides of the formula.**

QUESTIONS

QUICK TEST

1. If $a = \frac{3}{5}$ and $b = -2$, find the value of these expressions, giving your answer to 3 significant figures where appropriate.
   a. $ab - 5$
   b. $a^2 + b^2$
   c. $3a - 6ab$
2. Make u the subject of the formula:
$v^2 = u^2 + 2as$

EXAM PRACTICE

1. Make p the subject of the formula:
$5a - b = 3p + 2b$ [3 marks]
2. A person's body mass index (BMI), b, is calculated using the formula $b = \frac{m}{h^2}$ where m is the person's mass in kilograms and h is their height in metres.

   A person is classed as overweight if their BMI is greater than 25.

   Dan has a height of 179 cm and a mass of 84.5 kg. Would Dan be classed as overweight? You must show working to justify your answer. [3 marks]

Brackets and Factorisation

Multiplying out brackets helps to simplify algebraic expressions.

Expanding Single Brackets

Each term outside the bracket is multiplied by each separate term inside the bracket.

$5(x + 6) = 5x + 30$

Examples
Expand and simplify:

1. $-2(2x + 4) = \mathbf{-4x - 8}$
2. $5(2x - 3) = \mathbf{10x - 15}$
3. $8(x + 3) + 2(x - 1)$ ← Multiply out the brackets.
   $= 8x + 24 + 2x - 2$ ← Collect like terms.
   $= \mathbf{10x + 22}$
4. $3(2x - 5) - 2(x - 3)$ ← Multiply out the brackets.
   $= 6x - 15 - 2x + 6$ ← Collect like terms.
   $= \mathbf{4x - 9}$

Expanding Two Brackets

Every term in the second bracket must be multiplied by every term in the first bracket.

Often, but not always, the two middle terms are like terms and can be collected together.

$(x + 4)(x + 2) = x^2 + 2x + 4x + 8$
$= x^2 + 6x + 8$

Examples
Expand and simplify:

1. $(x + 4)(2x - 5) = 2x^2 - 5x + 8x - 20$
   $= \mathbf{2x^2 + 3x - 20}$
2. $(2x + 1)^2 = (2x + 1)(2x + 1)$
   $= 4x^2 + 2x + 2x + 1$
   $= \mathbf{4x^2 + 4x + 1}$

   Remember that x^2 means x multiplied by itself.
3. $(3x - 1)(x - 2) = 3x^2 - 6x - x + 2$
   $= \mathbf{3x^2 - 7x + 2}$
4. $(x - 4)(3x + 1) = 3x^2 + x - 12x - 4$
   $= \mathbf{3x^2 - 11x - 4}$
5. $(2x + 3y)(x - 2y) = 2x^2 - 4xy + 3xy - 6y^2$
   $= \mathbf{2x^2 - xy - 6y^2}$

Factorisation

Factorisation involves putting an expression into brackets.

One Bracket

$4x + 6 = 2(2x + 3)$

To factorise $4x + 6$:

- Recognise that 2 is the HCF of 4 and 6.
- Take out the highest common factor.
- The expression is completed inside the bracket so that when multiplied out it is equivalent to $4x + 6$.

Two Brackets

Two brackets are obtained when a quadratic expression of the type $ax^2 + bx + c$ is factorised.

Examples

1. $x^2 + 4x + 3 = \mathbf{(x + 1)(x + 3)}$
2. $x^2 - 7x + 12 = \mathbf{(x - 3)(x - 4)}$
3. $x^2 + 3x - 10 = \mathbf{(x + 5)(x - 2)}$
4. $x^2 - 64 = \mathbf{(x - 8)(x + 8)}$
5. $81x^2 - 25y^2 = \mathbf{(9x - 5y)(9x + 5y)}$
6. Here is a right-angled triangle.

   Four triangles are joined to enclose the square $MNPQ$. Work out the area of the square.

This is known as the 'difference of two squares'. In general, $x^2 - a^2 = (x - a)(x + a)$.

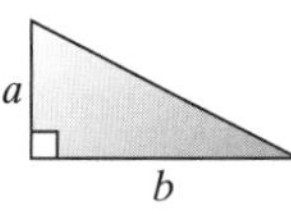

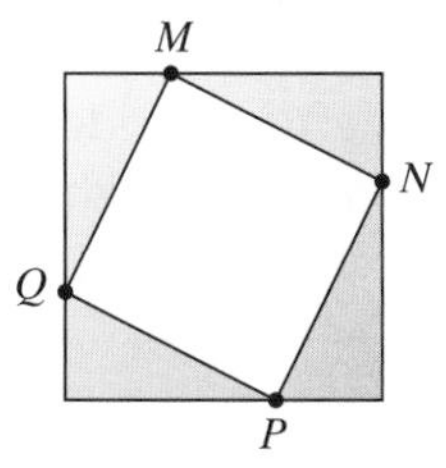

Area of the large square $= (a + b)^2$

Area of four triangles $= \dfrac{b \times h}{2} \times 4$

$= \dfrac{a \times b}{2} \times 4 = 2ab$

Area of square $MNPQ = (a + b)^2 - 2ab$

$= (a + b)(a + b) - 2ab$

$= a^2 + 2ab + b^2 - 2ab$

$= \mathbf{a^2 + b^2}$

SUMMARY

- **To expand single brackets, multiply each term outside the bracket by everything inside the bracket.**
- **To expand two brackets, multiply each term in the second bracket by each term in the first bracket.**
- **Factorising means putting in brackets. Look for the highest common factor (HCF).**

QUESTIONS

QUICK TEST

1. Expand and simplify:

   a. $(x + 3)(x - 2)$ **b.** $4x(x - 3)$

   c. $(x - 3)^2$

2. Factorise:

   a. $12xy - 6x^2$ **b.** $3a^2b + 6ab^2$

   c. $x^2 + 4x + 4$ **d.** $x^2 - 4x - 5$

   e. $x^2 - 100$

EXAM PRACTICE

1. Expand and simplify:

   a. $t(3t - 4)$ [1 mark]

   b. $4(2x - 1) - 2(x - 4)$ [2 marks]

2. Factorise:

   a. $y^2 + y$ [1 mark]

   b. $5p^2q - 10pq^2$ [2 marks]

   c. $(a + b)^2 + 4(a + b)$ [1 mark]

   d. $x^2 - 5x + 6$ [2 marks]

3. Show that $(a + b)^2 - 2b(a + b) = (a - b)(a + b)$ [3 marks]

DAY 2  15 Minutes

Equations 1

Equations involve an unknown value that needs to be worked out.

Equations need to be kept balanced, so whatever is done to one side of the equation (for example, adding) also needs to be done to the other side.

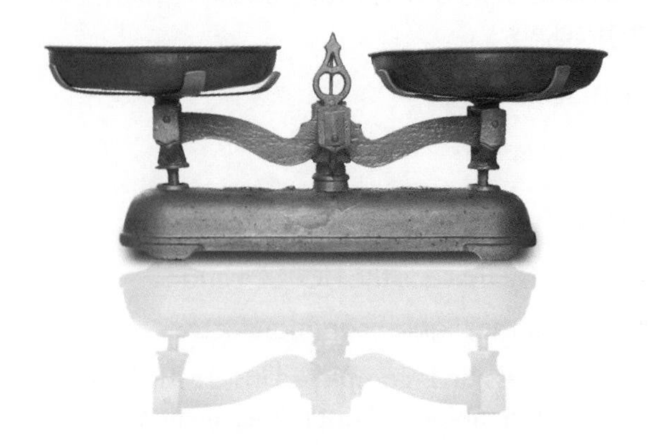

Linear Equations of the Form $ax + b = c$

Examples

1. Solve: $3x = 15$

$x = \frac{15}{3}$ ← Divide both sides by 3.

$x = \mathbf{5}$

2. Solve: $\frac{x}{3} = 6$

$x = 6 \times 3$ ← Multiply both sides by 3.

$x = \mathbf{18}$

3. Solve: $5x - 2 = 13$

$5x = 13 + 2$ ← Add 2 to both sides.

$5x = 15$

$x = \frac{15}{5}$ ← Divide both sides by 5.

$x = \mathbf{3}$

4. Solve: $3x + 1 = 13$

$3x = 13 - 1$ ← Subtract 1 from both sides.

$3x = 12$

$x = \frac{12}{3}$ ← Divide both sides by 3.

$x = \mathbf{4}$

5. Solve: $\frac{x}{6} - 1 = 3$

$\frac{x}{6} = 3 + 1$ ← Add 1 to both sides.

$\frac{x}{6} = 4$

$x = 4 \times 6$ ← Multiply both sides by 6.

$x = \mathbf{24}$

Linear Equations of the Form $ax + b = cx + d$

Examples

1. Solve:

$7x - 4 = 3x + 8$

$7x = 3x + 12$ ← Add 4 to both sides.

$4x = 12$ ← Subtract $3x$ from both sides.

$x = \frac{12}{4}$

$x = \mathbf{3}$

Check by substituting 3 into both sides of the equation:

$7 \times 3 - 4 = 17$

$3 \times 3 + 8 = 17$

Since both the left-hand side of the equation and the right-hand side of the equation give the same answer, $x = 3$ is correct ✔

2. Solve:

$5x + 3 = 2x - 5$

$5x = 2x - 5 - 3$ ← Subtract 3 from both sides.

$5x = 2x - 8$

$5x - 2x = -8$ ← Subtract $2x$ from both sides.

$3x = -8$

$x = -\frac{8}{3}$

$x = \mathbf{-2\frac{2}{3}}$

Linear Equations with Brackets

Examples

1. Solve:

$$5(x-1) = 3(x+2)$$

$$5x - 5 = 3x + 6$$

Just multiply out the brackets, then solve as normal.

$$5x = 3x + 11$$

$$2x = 11$$

$$x = \frac{11}{2}$$

$$x = \mathbf{5.5}$$

2. Solve:

$$5(2x+3) = 2(x-6)$$

$$10x + 15 = 2x - 12$$

$$10x = 2x - 12 - 15$$

$$10x = 2x - 27$$

$$10x - 2x = -27$$

$$8x = -27$$

$$x = -\frac{27}{8}$$

$$x = \mathbf{-3\frac{3}{8}}$$

3. Solve:

$$\frac{3(2x-1)}{5} = 6$$

$$3(2x-1) = 6 \times 5$$

Multiply both sides by 5.

$$6x - 3 = 30$$

$$6x = 33$$

$$x = \frac{33}{6}$$

$$x = \mathbf{5.5}$$

SUMMARY

- **Whatever you do to one side of an equation, do to the other side as well.**
- **Work through the solution step by step.**
- **Check by substituting in your answer.**

QUESTIONS

QUICK TEST

Solve the following equations:

1. $2x - 6 = 10$

2. $5 - 3x = 20$

3. $4(2 - 2x) = 12$

4. $6x + 3 = 2x - 10$

5. $7x - 4 = 3x - 6$

6. $5(x + 1) = 3(2x - 4)$

EXAM PRACTICE

1. Solve the equations:

a. $5x - 3 = 9$ [2 marks]

b. $7x + 4 = 3x - 6$ [3 marks]

c. $3(4y - 1) = 21$ [3 marks]

2. Solve:

a. $5 - 2x = 3(x + 2)$ [3 marks]

b. $\frac{3x - 1}{3} = 4 + 2x$ [3 marks]

3. Joe is asked to solve the equation $3(x - 6) = 42$

Here is his working:

$$3(x-6) = 42$$

$$3x - 6 = 42$$

$$3x = 42 + 6$$

$$3x = 48$$

$$x = 16$$

What mistake did he make? [1 mark]

Equations 2

Equations and Identities

An **equation** can be solved to find an unknown quantity.

$5x + 2 = 12$ can be solved to find $x = 2$

An **identity** is true for all values of x.
$(x + 5)^2$ is identically equal to $x^2 + 10x + 25$.

You can write this as $(x + 5)^2 \equiv x^2 + 10x + 25$

Equation Problems

When solving equation problems, the first step is to write down the information that you know.

Example
The perimeter of this rectangle is 30 cm.

Work out the value of y and find the length of the rectangle.

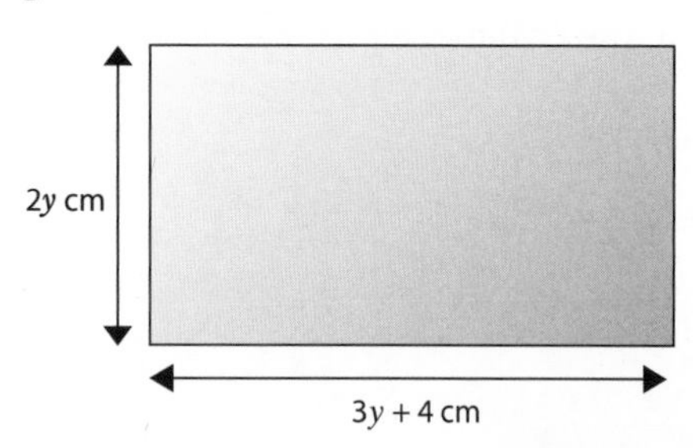

$3y + 4 + 2y + 3y + 4 + 2y = 30$ ← Write down what you know.

$10y + 8 = 30$

$10y = 30 - 8$ ← Simplify the expression and solve as normal.

$10y = 22$

$y = \mathbf{2.2}$

Length of rectangle $= 3 \times 2.2 + 4$

$= \mathbf{10.6\ cm}$

Equations and Fractions

Examples

1. Solve $\frac{x+4}{3} = 10$

$x + 4 = 30$ ← Multiply both sides by 3.

$x = \mathbf{26}$ ← Subtract 4 from both sides.

2. Solve $\frac{x+2}{3} + \frac{x-1}{2} = \frac{15}{6}$

$2(x + 2) + 3(x - 1) = 15$ ← 6 is the lowest common multiple of 2, 3 and 6. Multiply both sides of the equation by 6.

$2x + 4 + 3x - 3 = 15$ ← Expand the brackets.

$5x + 1 = 15$ ← Solve as normal.

$5x = 15 - 1$

$x = \mathbf{\frac{14}{5}}$ or $\mathbf{2\frac{4}{5}}$

Solving Quadratic Equations

A **quadratic equation** can be solved by factorising. Firstly, check if it is written in the form $ax^2 + bx + c = 0$.

Solve the equation:

$$x^2 - x - 6 = 0$$

Need to factorise into two brackets $(x \pm ?)(x \pm ?) = 0$

Must check that the equation equals zero

- Factorise into two brackets: $(x + 2)(x - 3) = 0$
- Since the equation equals zero, one of the brackets must equal zero.

  either $(x + 2) = 0$ or $(x - 3) = 0$

  So $x = -2$ or $x = 3$

Example

Solve $x^2 - 7x + 10 = 0$

$(x - 2)(x - 5) = 0$

either $(x - 2) = 0$ or $(x - 5) = 0$

so $x = \mathbf{2}$ or $x = \mathbf{5}$

Using Graphs to Solve Quadratic Equations

Graphs can be used to solve equations by looking at their points of intersection with the x-axis or with each other.

Example

The graph shows the equation $y = x^2 - 8x + 7$

Use the graph to solve $x^2 - 8x + 7 = 0$

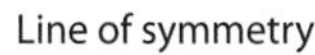

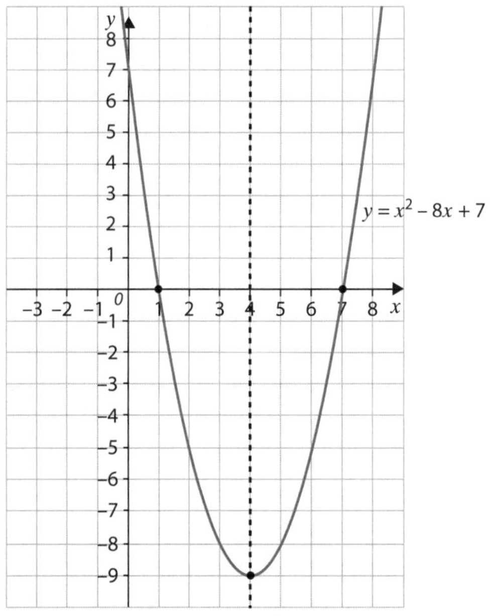

$x^2 - 8x + 7 = 0$ is where the curve crosses the x-axis, so the solutions are $\boldsymbol{x = 1}$, $\boldsymbol{x = 7}$. These solutions are called the **roots** of the equation.

SUMMARY

- **When solving equation problems, write down what you know, simplify and then solve as normal.**
- **Quadratic equations can be solved by factorisation.**
- **Graphs can be used to solve equations by looking at their points of intersection with the x-axis. These values of x are called the roots of the equation.**

QUESTIONS

QUICK TEST

1. The perimeter of this triangle is 60 cm. Work out the value of x and find the shortest length.

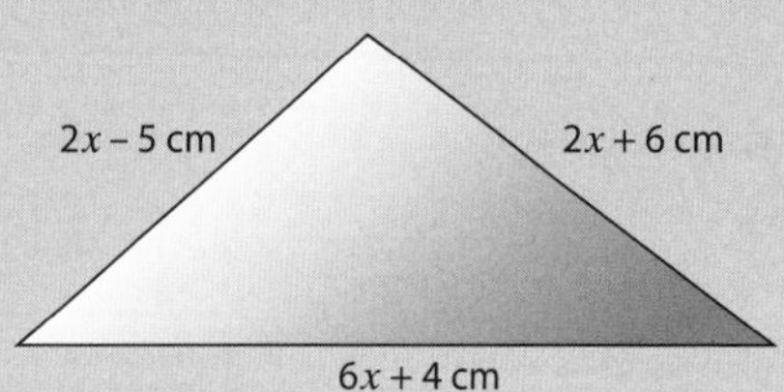

2. Solve:
   a. $x^2 - 7x = 0$
   b. $x^2 + 8x + 15 = 0$
   c. $x^2 - 5x + 6 = 0$

EXAM PRACTICE

1. The sizes of the angles, in degrees, of the quadrilateral are:

   $x + 30°$
   $2x$
   $x + 50°$
   $x + 10°$

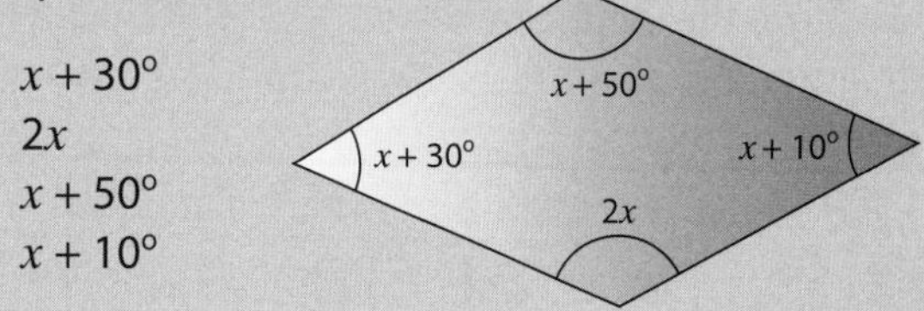

   Work out the smallest angle of the quadrilateral. [4 marks]

2. Solve $\frac{x-2}{4} + \frac{x+3}{2} = \frac{9}{4}$ [3 marks]

Simultaneous Linear Equations

Two equations with two unknowns are called **simultaneous equations**. They can be solved algebraically or graphically.

Solving Algebraically (Elimination Method)

Solve simultaneously:

$3x + 2y = 8$
$2x - 3y = 14$

Label the equations ① and ②.

$3x + 2y = 8$ ①
$2x - 3y = 14$ ②

Since no coefficients match, multiply equation ① by 2 and equation ② by 3.

$6x + 4y = 16$
$6x - 9y = 42$

Rename them equations ③ and ④.

$6x + 4y = 16$ ③
$6x - 9y = 42$ ④

The coefficient of x in equations ③ and ④ is the same. Subtract equation ④ from equation ③ and solve to find y.

$0x + 13y = -26$
$y = -26 \div 13$
$y = -2$

Note $4y - (-9y)$
$= 4y + 9y$
$= 13y$

Substitute the value of $y = -2$ into equation ①. Solve this equation to find x.

$3x + 2 \times (-2) = 8$
$3x + (-4) = 8$
$3x = 8 + 4$
$3x = 12$
$x = 4$

You could substitute into ②.

Check in equation ②.

$(2 \times 4) - (3 \times -2) = 14$ ✓

Solution is: $\boldsymbol{x = 4, y = -2}$

Solving Graphically

The point at which any two graphs **intersect** represents the simultaneous solution of their equations.

Example

Solve the simultaneous equations:

$2x + 3y = 6$

$x + y = 1$

Draw the graph of:

$2x + 3y = 6$

When $x = 0, 3y = 6 \therefore y = 2$ (0, 2)

When $y = 0, 2x = 6 \therefore x = 3$ (3, 0)

Draw the graph of:

$x + y = 1$

When $x = 0, y = 1$ (0, 1)

When $y = 0, x = 1$ (1, 0)

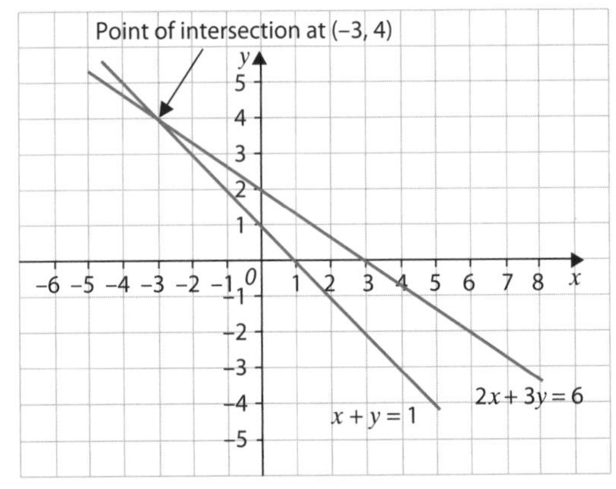

At the point of intersection: $\boldsymbol{x = -3}$, $\boldsymbol{y = 4}$. This is the solution of the simultaneous equations.

SUMMARY

- **When you are solving simultaneous equations, there are two equations with two unknowns.**
- **The elimination method involves eliminating one of the variables.**
- **The point where two graphs cross gives the solution to both equations.**

QUESTIONS

QUICK TEST

1. Solve the simultaneous equations:

   $4b + 7a = 10$

   $2b + 3a = 3$

2. The diagram shows the graphs of the lines:

   $x + y = 6$ and $y = x + 2$

   Use the diagram to solve the simultaneous equations $x + y = 6$ and $y = x + 2$.

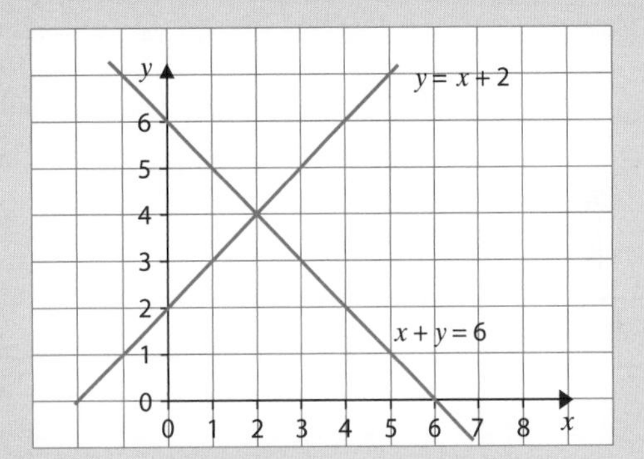

EXAM PRACTICE

1. Solve the simultaneous equations:

   $5a - 2b = 19$

   $3a + 4b = 1$ [4 marks]

2. Frances and Patrick were organising a children's party. They went to the toy shop and bought some hats and balloons. Frances bought five hats and four balloons. She paid £22. Patrick bought three hats and five balloons. He paid £21.

   The cost of a hat was x pounds. The cost of a balloon was y pounds.

   Work out the cost of one hat and the cost of one balloon. [5 marks]

Sequences

A **sequence** is a set of numbers that follow a particular rule. The word **'term'** is often used to describe each number in the sequence.

A term-to-term sequence means you can find a rule for each term based on the previous term in the sequence. For example, in the sequence 2, 4, 6, 8 … you add 2 each time to go from one term to the next.

Special Sequences

Odd numbers 1, 3, 5, 7, 9 … nth term is $2n - 1$

Even numbers 2, 4, 6, 8, 10 … nth term is $2n$

Square Numbers

1 4 9 16 25 … nth term is n^2

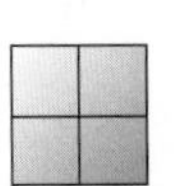
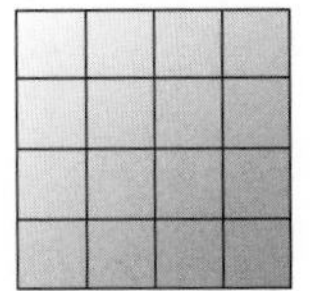

$1^2 = 1 \times 1$ $2^2 = 2 \times 2$ $3^2 = 3 \times 3$ $4^2 = 4 \times 4$ $5^2 = 5 \times 5$

You need to know the square numbers up to 15×15.

Cube Numbers

1 8 27 64 125 …

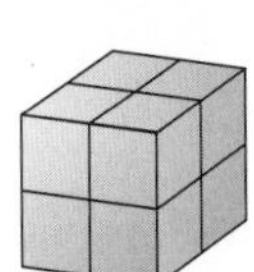
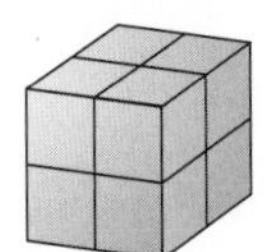

$1^3 = 1 \times 1 \times 1$ $2^3 = 2 \times 2 \times 2$ $3^3 = 3 \times 3 \times 3$

Triangular Numbers

1 3 6 10 15 …

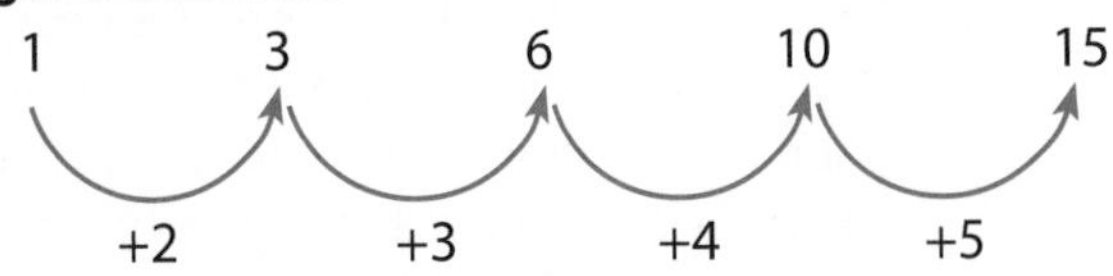

Fibonacci Sequence

1, 1, 2, 3, 5, 8, 13 … Add the previous two terms.

Another type of Fibonacci sequence is the Lucas series: 2, 1, 3, 4, 7, 11, 18 …

Function Machines and Mapping

Function machines are useful when finding a relationship between two **variables**. For example, when numbers are fed into this machine they are first multiplied by 2 and 1 is then added:

Input (x) ⟶ $\times 2 + 1$ ⟶ Output (y)

- If 1 is fed in, 3 comes out ($1 \times 2 + 1 = 3$).
- If 2 is fed in, 5 comes out ($2 \times 2 + 1 = 5$), etc.

Finding the nth Term of an Arithmetic Sequence

The nth term is the rule for a sequence and is often denoted by U_n. For example, the 8th term is U_8.

For a **linear** or **arithmetic** sequence, the nth term takes the form:

$U_n = an + b$

Example

Find the nth term of this sequence: 2, 6, 10, 14…

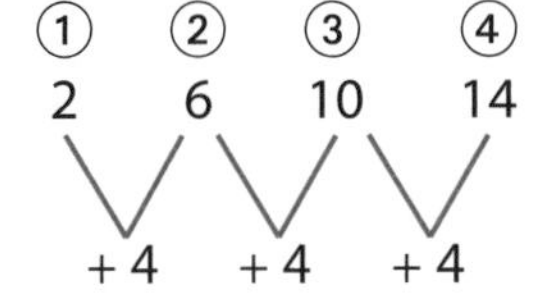

- See how the numbers are jumping (going up in 4s).
- The nth term is $4n$ + or – something.
- Try $4n$ on the first term. This gives $4 \times 1 = 4$, but the first term is 2 … so subtract 2.
- The rule is $4n - 2$
- Test this rule on the other terms:

  $1 \rightarrow (4 \times 1) - 2 = 2$
  $2 \rightarrow (4 \times 2) - 2 = 6$
  $3 \rightarrow (4 \times 3) - 2 = 10$
  It works on all of them.
- nth term is $\mathbf{4n - 2}$
  The 20th term in the sequence would be:
  $(4 \times 20) - 2 = 78$

Geometric Sequences

In a **geometric sequence**, you multiply by a constant number (**common ratio**), r, to go from one term to the next.

×3 ×3 ×3 ×3
2 6 18 54 162

Each term is multiplied by 3 each time.
So the common ratio, $r = 3$

The term-to-term formula is $U_{n+1} = 3U_n$, with $U_1 = 2$

The position-to-term formula is $U_n = 2 \times 3^{n-1}$

SUMMARY

- **A sequence is a set of numbers that follow a particular rule.**
- **The nth term is the rule for a sequence and is denoted by U_n.**
- **Given a sequence, you will need to be able to work out the nth term.**

QUESTIONS

QUICK TEST

1. The cards show the nth term of some sequences:

$2n$	$4n + 1$	$3n + 2$	$5n - 1$	$2 - n$

   Match the cards with the sequences below:

   a. 5, 9, 13, 17 …

   b. 1, 0, –1, –2 …

   c. 2, 4, 6, 8, 10 …

   d. 5, 8, 11, 14, 17 …

   e. 4, 9, 14, 19 …

2. **a.** Write down the next two terms in the sequence below:

   2 8 32 128 512

   b. Explain how to find the next number in the sequence.

EXAM PRACTICE

1. Here are the first four terms of an arithmetic sequence:

   5 7 9 11

   Find an expression in terms of n for the nth term of the sequence. [2 marks]

2. The nth term of a sequence is $2n^2 + 1$. Chloe says that 101 is a number in the sequence.

   Explain whether Chloe is correct. [2 marks]

Inequalities

An **inequality** is a statement showing two quantities that are not equal.

Inequalities can be solved in exactly the same way as equations, except that when multiplying or dividing by a negative number you must reverse the inequality sign.

The Inequality Symbols

$>$ means **greater than**

$<$ means **less than**

$\geqslant$ means **greater than or equal to**

$\leqslant$ means **less than or equal to**

Examples

1. Solve $2x - 2 < 10$

$2x < 10 + 2$

$2x < 12$

$\mathbf{x < 6}$

2. Solve $4x - 2 \geqslant 6$

$4x \geqslant 6 + 2$ ← Add 2 to both sides.

$4x \geqslant 8$

$x \geqslant \frac{8}{4}$ ← Divide both sides by 4.

$\mathbf{x \geqslant 2}$

3. Solve $3 - 2x \geqslant 9$

$-2x \geqslant 9 - 3$

$-2x \geqslant 6$

$x \leqslant \frac{6}{-2}$ ← Divide by –2 and reverse the inequality.

$\mathbf{x \leqslant -3}$

Inequalities Involving Fractions

If the inequality involves fractions, you should multiply through in order to remove the fractions.

Example

Solve the inequality $\frac{3x - 2}{4} > 4$

$3x - 2 > 16$ ← Multiply both sides by 4.

$3x > 16 + 2$ ← Add 2 to both sides.

$3x > 18$

$x > \frac{18}{3}$ ← Divide both sides by 3.

$\mathbf{x > 6}$

Two Inequalities

When there are two inequalities, it is important to do the same thing to all parts of the inequality.

Examples

1. Solve $-7 < 3x - 1 \leqslant 11$

$-6 < 3x \leqslant 12$ ← Add 1 to each part of the inequality.

$\mathbf{-2 < x \leqslant 4}$ ← Divide each part of the inequality by 3.

The **integer** values that satisfy this inequality are –1, 0, 1, 2, 3, 4.

2. Solve $2 < \frac{2x - 5}{3} < 5$

$6 < 2x - 5 < 15$ ← Multiply each part of the inequality by 3.

$11 < 2x < 20$ ← Add 5 to each part of the inequality.

$\mathbf{5.5 < x < 10}$ ← Divide each part of the inequality by 2.

The integer values that satisfy this inequality are 6, 7, 8, 9.

Number Lines

Inequalities can be shown on a number line.

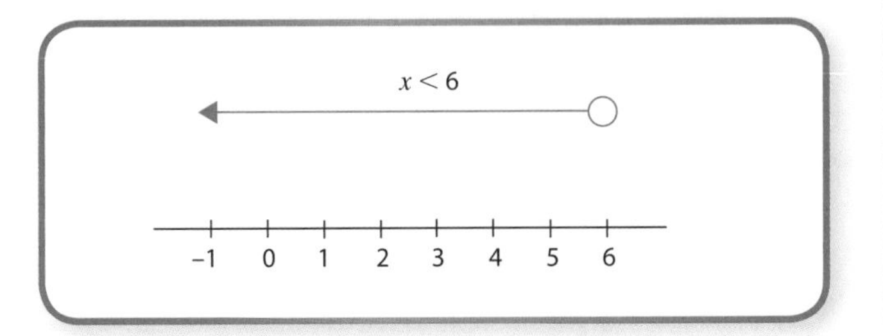

The open circle means that 6 is not included.

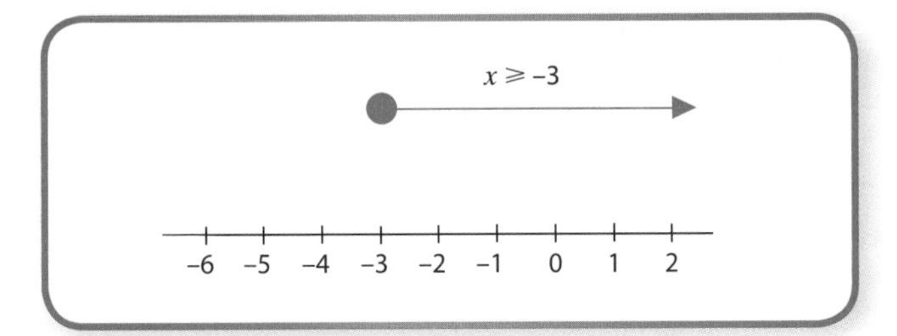

The solid circle means that −3 is included.

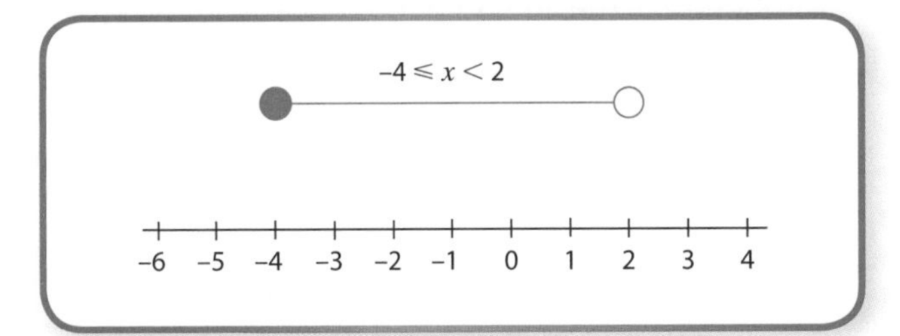

The integer values that satisfy this inequality are −4, −3, −2, −1, 0, 1.

SUMMARY

- **$>$ greater than**

  $<$ less than

  $\geqslant$ greater than or equal to

  $\leqslant$ less than or equal to
- **On number lines, an open circle means the value is not included in the inequality and a solid circle means the value is included in the inequality.**

QUESTIONS

QUICK TEST

1. Solve the following inequalities:

a. $5x - 1 < 10$

b. $6 \leqslant 3x + 2 < 11$

c. $3 - 5x < 12$

EXAM PRACTICE

1. n is an integer such that $-6 < 2n \leqslant 8$.

List all the possible values of n. [1 mark]

2. Solve the inequality $4 + x > 7x - 8$. [2 marks]

3. a. Solve the inequality $\frac{3x+5}{4} \leqslant 5$ [3 marks]

b. On the number line below, represent the solutions of the inequality $\frac{3x+5}{4} \leqslant 5$ [1 mark]

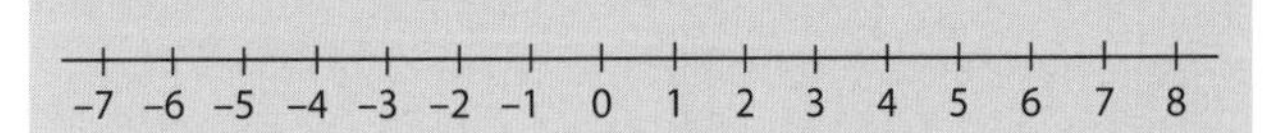

Straight-line Graphs

Drawing Straight-line Graphs

The general equation of a straight line is:

$$y = mx + c$$

m is the **gradient**. c is the **intercept** on the y-axis.

To draw the graph of $y = 3x - 4$:

- Work out the coordinates of the points that lie on the line $y = 3x - 4$ by drawing a table of values for x.

  Substitute the x values into the equation $y = 3x - 4$, to find the values of y
  e.g. $x = 2$, $y = 3 \times 2 - 4 = 2$

x	–1	0	2	4
y	–7	–4	2	8

- The coordinates of the points on the line are:
  (–1, –7) (0, –4) (2, 2) (4, 8)
  Just read them from the table of values.
- Plot the points (remember to read across first, then up/down) and join with a straight line.
- The line $y = 3x - 4$ is drawn.

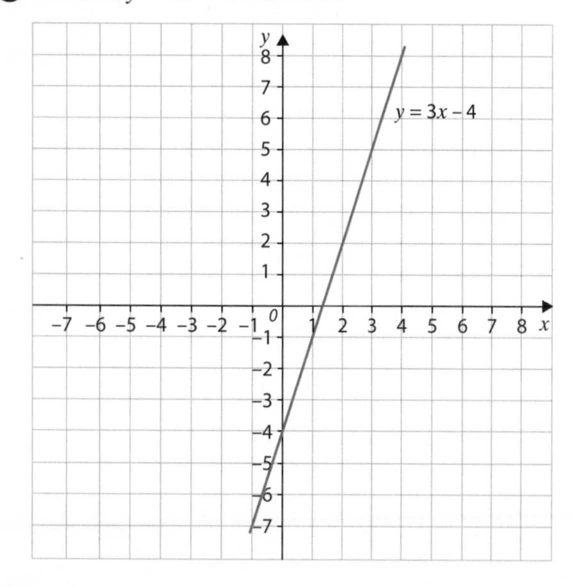

- Label the line once you have drawn it.

Gradient of a Straight Line

Be careful when finding the gradient: double-check the scales.

$$\text{Gradient} = \frac{\text{change in } y}{\text{change in } x}$$

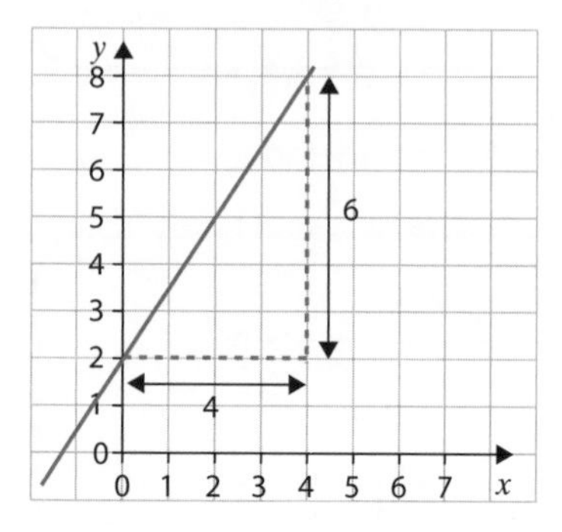

For the straight line above:

$$\text{Gradient} = \frac{6}{4} = \frac{3}{2} = 1.5$$

The equation of the line is $y = 1.5x + 2$

Parallel lines have the same gradients:

$y = 2x - 3$ and $y = 2x + 5$ are parallel with a gradient of 2.

Positive and Negative Gradients

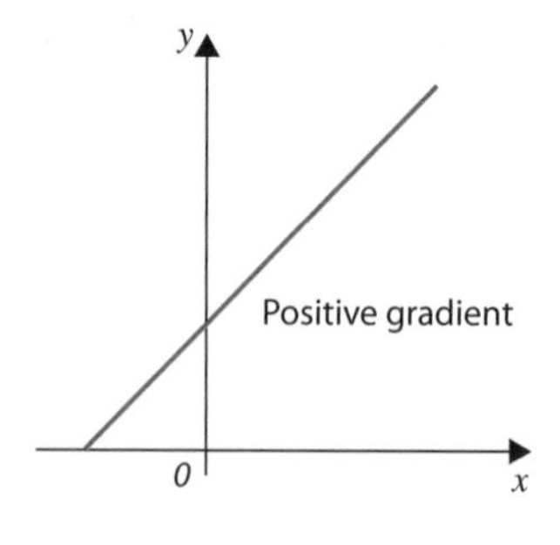

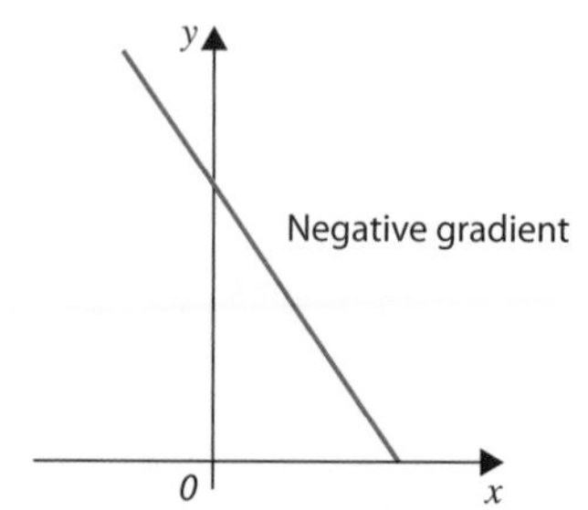

Finding the Equations of Straight Lines

You need to be able to find the equations of lines when key information is given. Remember the equation of a straight line is in the form $y = mx + c$.

Example

Find the equation of a line with gradient –3 going through the point (–2, 10).

The equation of the line is $y = mx + c$

$10 = (-3 \times -2) + c$

$10 = 6 + c$

$c = 4$

Substitute the gradient $m = -3$ and the coordinates into the equation; use $x = -2$ and $y = 10$

Equation of the line is $\mathbf{y = -3x + 4}$

Finding the Midpoint of a Line Segment

The midpoint of a line segment between two points can be worked out by finding the mean of the x coordinates and the mean of the y coordinates of the points.

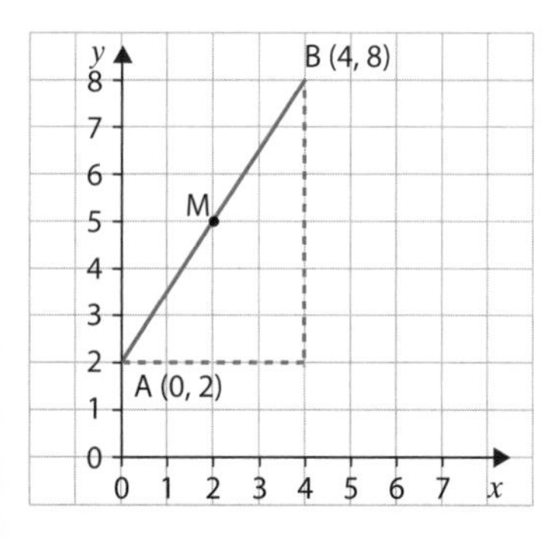

The midpoint of a line that joins the point A(x_1, y_1) and B(x_2, y_2) is:

$$\left(\frac{(x_1 + x_2)}{2}, \frac{(y_1 + y_2)}{2}\right)$$

The midpoint, M, of the line AB drawn here is:

$$\left(\frac{(0 + 4)}{2}, \frac{(2 + 8)}{2}\right) = (2, 5)$$

SUMMARY

- **The general equation for a straight line is $y = mx + c$**
- **Gradient = $\frac{\text{change in } y}{\text{change in } x}$**
- **Work out the midpoint of a line segment by finding the mean of the x coordinates and the mean of the y coordinates.**
- **Parallel lines have the same gradient.**

QUESTIONS

QUICK TEST

1. **a.** Complete the table of values for $y = 2x + 3$.

x	–2	–1	0	1	2	3
y						

b. Draw the graph of $y = 2x + 3$.

EXAM PRACTICE

1.

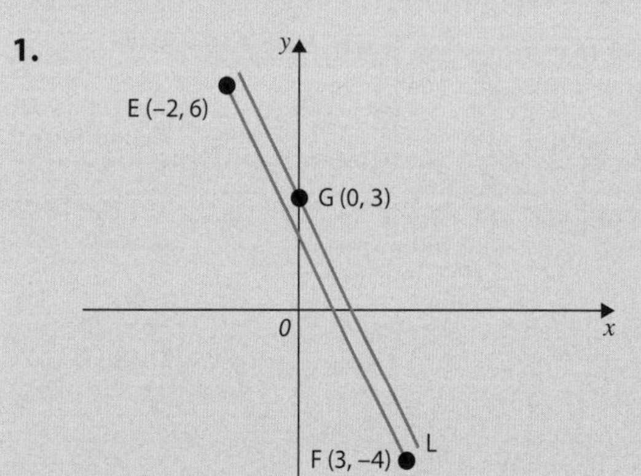

The diagram shows three points:

E (–2, 6) F (3, –4) G (0, 3)

Line L is parallel to EF and passes through G.

a. Find an equation for the line L. [3 marks]

b. Find the midpoint of the line EF. [2 marks]

c. Write down the equation of a line which goes through the point G and is not parallel to L. [2 marks]

Curved Graphs

Quadratic Graphs

Quadratic graphs are of the form $y = ax^2 + bx + c$ where $a \neq 0$. Quadratic graphs have an x^2 term as the highest power of x.

They will be ∪ shaped if the **coefficient** of x^2 is positive, and ∩ shaped if the coefficient of x^2 is negative.

To draw the graph of $y = x^2 - 2x - 6$, using values of x from –2 to 4:

- Draw a table of values.

  Fill in the table of values by substituting the values of x into the equation.

  e.g. $x = 1, \quad y = 1^2 - 2 \times 1 - 6 = -7$

  Coordinates are (1, –7)

x	–2	–1	0	1	2	3	4
y	2	–3	–6	–7	–6	–3	2

- Draw the axes on graph paper and plot the points.
- Join the points with a smooth curve.
- Label the curve.

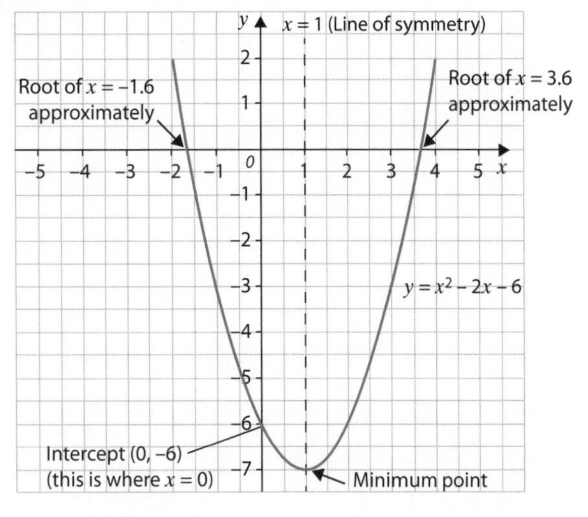

- The **minimum** point is (1, –7). This is also known as a **turning point.** Every quadratic equation has a turning point at a minimum (if $a > 0$) or a **maximum** point (if $a < 0$).
- The **line of symmetry** is at $x = 1$. The line of symmetry occurs at $x = -\frac{b}{2a}$. For the example above, $b = -2$ and $a = 1$. So $x = -\frac{-2}{2 \times 1}$, hence $x = 1$.

Graph Shapes

$y = mx + c$

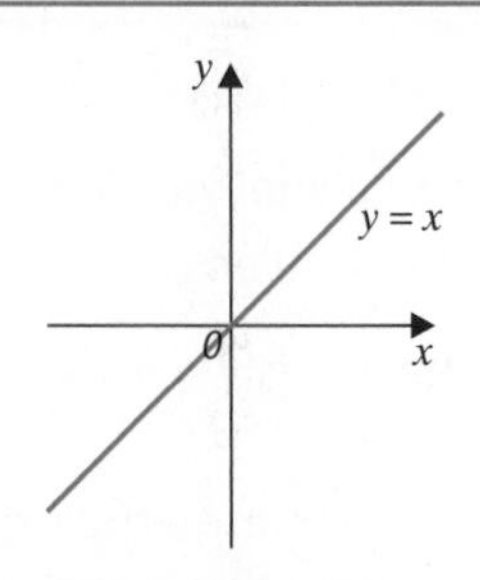

$y = ax^2 + bx + c$

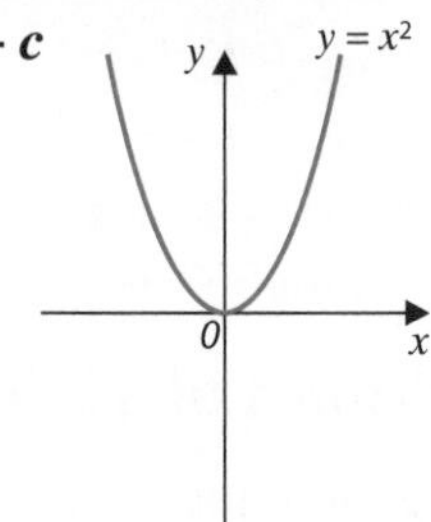

$y = \frac{k}{x}$

(Reciprocal functions)

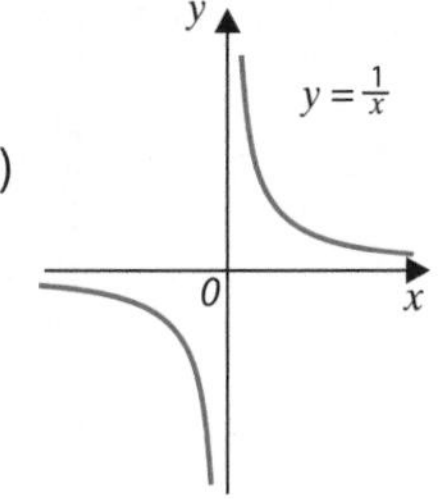

$y = x^3$

(Cubic functions)

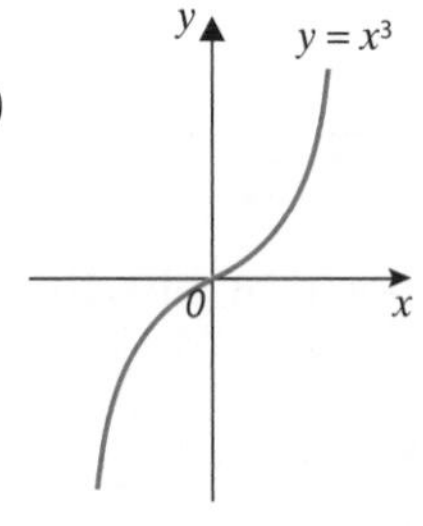

$y = ax^3 + bx^2 + cx + d$

(Cubic functions)

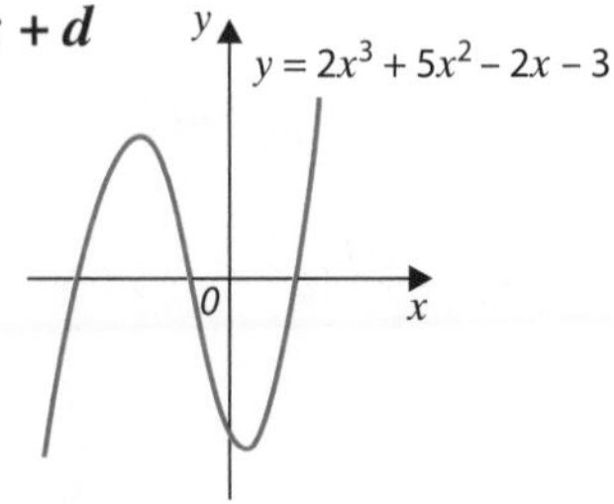

Example

Match each graph to one of the following equations.

$y = x^2 - 4$ $y = 5 - 2x$ $y = x^3$ $y = 3x - 1$

Graph A

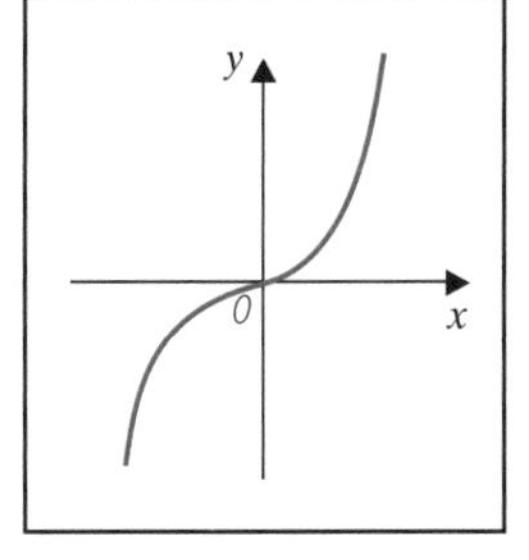

Graph B

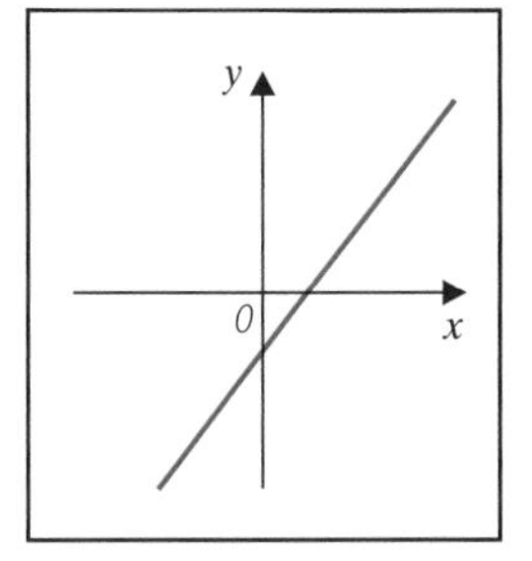

Graph C

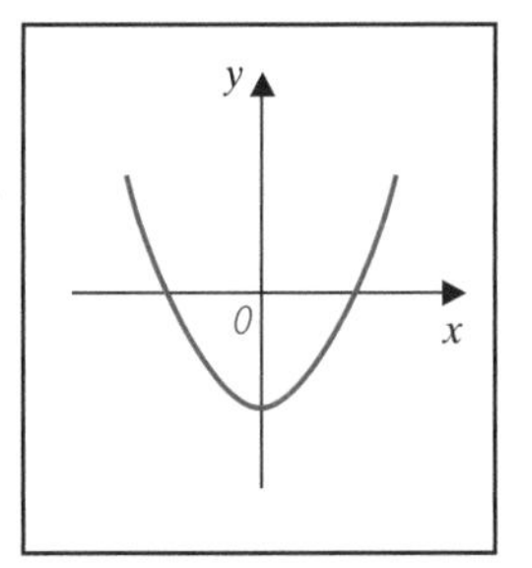

Graph D

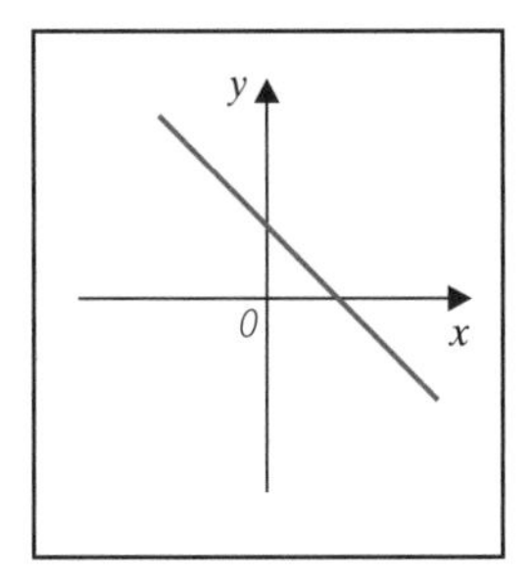

Graph A is $y = x^3$

Graph B is $y = 3x - 1$

Graph C is $y = x^2 - 4$

Graph D is $y = 5 - 2x$

SUMMARY

- **Make sure you know the different graph shapes.**
- **Quadratic graphs have an x^2 term as the highest power of x. They are ∪ shaped or ∩ shaped.**
- **Cubic graphs have an x^3 term as the highest power of x.**
- **The reciprocal function is $y = \frac{k}{x}$**

QUESTIONS

QUICK TEST

1. Match each graph below to one of the equations.

$y = x^3 - 5$ $y = 2 - x^2$ $y = 4x + 2$ $y = \frac{3}{x}$

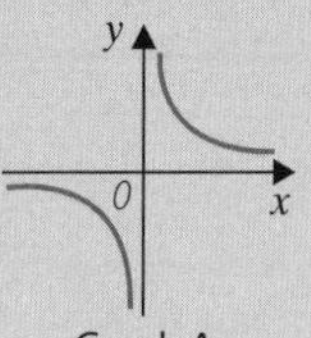

Graph A

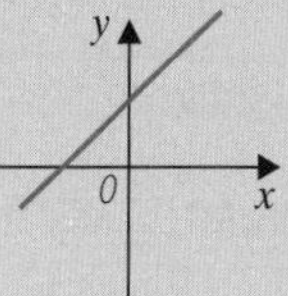

Graph B

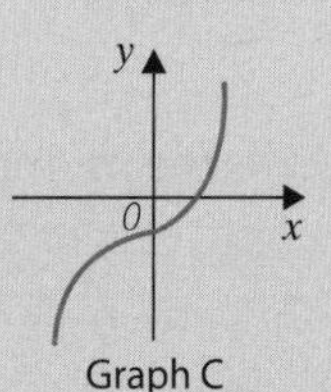

Graph C

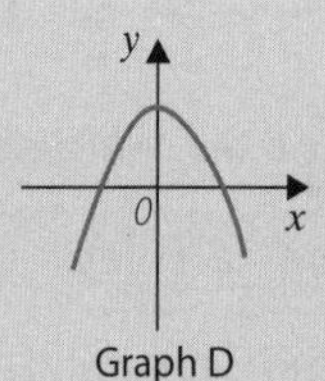

Graph D

EXAM PRACTICE

1. a. Complete the table of values for $y = x^3 - 1$.

x	−3	−2	−1	0	1	2	3
y				−1			

[2 marks]

b. Using a suitable scale, draw the graph of $y = x^3 - 1$. [3 marks]

c. From the graph find the approximate value of x when $y = 15$. [2 marks]

Percentages

Percentages can be expressed as fractions with a **denominator** of 100.

This is the percentage sign:

%

Percentage of a Quantity

OF means multiply. To find the percentage of a quantity, write the percentage as a fraction with a denominator of 100 and multiply by the quantity.

Example

Find 30% of 80 kg.

$\frac{30}{100} \times 80 = 24$ kg

On the calculator, key in:

=

$30\% = \frac{30}{100} = 0.3$, which is known as the **multiplier**.

For the **non-calculator** paper:

- find 10% by dividing by 10

  10% of 80 kg

  $= \frac{80}{10}$

  = 8 kg

- then multiply by 3 to get 30%

  3×8

  = **24 kg**

Increasing and Decreasing Quantities by a Percentage

Percentages appear in everyday life and you will often need to find the value of quantities after a percentage increase or decrease.

Examples

1. An MP3 player costs £165. In a sale it is reduced by 15%. Work out the cost of the MP3 player in the sale.

   Method 1

   15% of £165

   $= \frac{15}{100} \times 165$

   = £24.75

   Price of the MP3 player in the sale

   = £165 – £24.75 = **£140.25**

   Method 2 (using a multiplier)

   $1 - 0.15 = 0.85$

   0.85×165

   = **£140.25**

   0.85 is the multiplier since the price is going down.

2. Louisa works out the cost of her gas bill. At the start of a three-month period, the gas meter reading was 12 447 units. At the end of the three-month period, the gas meter reading was 12 721.

   Each unit of gas used costs 47p. VAT is charged at 5%. Work out the total cost of Louisa's gas bill.

   Units used 12 721 – 12 447
   = 274

   Cost of gas 274×47p
   = £128.78

   VAT at 5%
   Multiplier $1 + 0.05 = 1.05$
   1.05×128.78

   1.05 is the multiplier since the price is going up.

   Gas bill = **£135.22**

   Round to the nearest penny.

One Quantity as a Percentage of Another

To express one quantity as a percentage of another, divide the first quantity by the second quantity and multiply by 100%.

Example

Matthew got 46 out of 75 in a Science test. He got 65% in a Maths test. In which test did he do better?

Work out the percentage he got in the Science test.

$\frac{46}{75} \times 100\%$ ← Make a fraction and multiply by 100%.

$= \mathbf{61.\dot{3}\%}$

On the calculator, key in:

In the Maths test Matthew got 65%, so he did better in the Maths test than the Science test.

SUMMARY

- **Percentages can be expressed as fractions with a denominator of 100.**
- **To find the percentage of a quantity, write the percentage as a fraction with a denominator of 100 and multiply by the quantity.**
- **To express one quantity as a percentage of another, divide the first quantity by the second quantity and multiply by 100%.**
- **OF means multiply.**

QUESTIONS

QUICK TEST

1. Without using a calculator, work out the following:
   a. 12% of 50 kg **b.** 30% of £2000 **c.** 5% of £60 **d.** 35% of 720 g
2. Without using a calculator, change each fraction into a percentage:
   a. $\frac{16}{50}$ **b.** $\frac{46}{200}$ **c.** $\frac{15}{20}$ **d.** $\frac{21}{25}$
3. Reduce £225 by 20%.
4. Express 32 as a percentage of 40.

EXAM PRACTICE

1. Jonathan is buying a new television. He sees three different advertisements for the same television.

BEST TV SHOP	**DRYMONS**	**MARK'S ELECTRICALS**
Normal Price £556 SALE: $\frac{1}{5}$ off Normal Price	TV Normal Price £495 Sale 10% off	TV £385 PLUS VAT at 20%

   Jonathan wants to buy the cheapest television. From which shop should Jonathan purchase his television? You must show full working out and give a reason for your answer. [4 marks]

2. Madeleine earns £48 500 per year. She does not pay income tax on the first £10 600 of her salary (which is deducted before tax is calculated). She then pays the basic rate of tax at 20% on her earnings up to £31 786, and the higher rate of tax at 40% on anything over £31 786. How much tax will Madeleine pay? [5 marks]

Repeated Percentage Change

Percentage Change

$$\text{Percentage change} = \frac{\text{change}}{\text{original}} \times 100\%$$

Examples

1. Tammy bought a flat for £185 000. Three years later she sold it for £242 000. What is her percentage profit?

   Profit is £242 000 – £185 000

   = £57 000

   Percentage profit is $\frac{57\,000}{185\,000} \times 100\%$

   = **30.8%** (3 s.f.)

2. Jackie bought a car for £12 500 and sold it two years later for £7250. Work out her percentage loss.

   Loss is £12 500 – £7250

   = £5250

   Percentage loss is $\frac{5250}{12\,500} \times 100\%$

   = **42%**

Repeated Percentage Change

A quantity can increase or decrease in value each year by the same or a different percentage. These quantities will change in value at the end of each year. To calculate repeated percentage change, two methods are explained in the example below.

Example

A car was bought for £12 500. Each year it depreciated in value by 15%. What was the car worth after three years?

You must remember **not** to do 3 × 15% = 45% reduction over three years!

Method 1

- Find 100% – 15% = 85% of the value of the car first.

  Year 1: $\frac{85}{100} \times £12\,500 = £10\,625$

- Then work out the value year by year. (£10 625 depreciates in value by 15%.)

  Year 2: $\frac{85}{100} \times £10\,625 = £9031.25$

  (£9031.25 depreciates in value by 15%.)

  Year 3: $\frac{85}{100} \times £9031.25 =$ **£7676.56**

Method 2

- A quick way to work this out is by using a **multiplier**.
- Finding 85% of the value of the car is the same as multiplying by 0.85

  Year 1: 0.85 × £12 500 = £10 625

  Year 2: 0.85 × £10 625 = £9031.25

  Year 3: 0.85 × £9031.25 = **£7676.56**

- This is the same as working out $(0.85)^3 \times £12\,500 =$ **£7676.56**

Compound Interest

Compound interest is where the bank pays interest on the interest already earned as well as on the original money.

Example

Becky has £3200 in her savings account and compound interest is paid at 3.2% per annum. How much will she have in her account after four years?

100% + 3.2% = 103.2%

= 1.032 ← This is the multiplier.

Year 1: 1.032 × £3200 = £3302.40

Year 2: 1.032 × £3302.40 = £3408.08

Year 3: 1.032 × £3408.08 = £3517.14

Year 4: 1.032 × £3517.14 = £3629.68

Total = **£3629.68**

A quicker way is to multiply £3200 by $(1.032)^4$

Number of years

£3200 × $(1.032)^4$ = **£3629.68**

Original

Multiplier

Simple Interest

Simple interest is the interest paid each year. It is the same amount each year.

The simple interest on £3200 invested for four years at 3.2% per annum would be:

$\frac{3.2}{100} \times 3200 =$ £102.40 for one year

Interest over four years would be 4 × £102.40 = £409.60

Total in account after four years would be £3609.60

SUMMARY

- **A quick way to work out repeated percentage change is to use a multiplier.**
- **Compound interest is where the bank pays interest on the interest already earned as well as on the original money.**
- **Simple interest is the interest paid each year. It is the same amount each year.**

QUESTIONS

QUICK TEST

1. A car is bought for £8500. Two years later it is sold for £4105. Work out the percentage loss. Give your answer to 3 significant figures.
2. A flat was bought for £85 000 in 2013. The flat rose in value by 12% in 2014 and by 28% in 2015. How much was the flat worth at the end of 2015?

EXAM PRACTICE

1. Shamil invests £3000 in each of two bank accounts. The terms of the bank accounts are shown below.

Savvy Saver	**Money Grows**
Simple interest at 2.5% per annum.	Compound interest at 2.5% per annum.

Shamil says that he will earn the same amount of interest from both bank accounts in two years.

Decide whether Shamil is correct. You must show full working to justify your answer.

[3 marks]

Reverse Percentage Problems

In reverse percentage problems you are given the final amount after a **percentage increase** or **decrease**. You have to then find the value of the original quantity. These are quite tricky, so think carefully.

Example 1

The price of a television is reduced by 15% in the sales. It now costs £352.75

What was the original price?

- The sale price is 100% – 15% = 85% of the pre-sale price (x)
- 85% = 0.85 ← This is the **multiplier**.
- $0.85 \times x = £352.75$

$$x = \frac{£352.75}{0.85}$$

Original price was **£415**

Check:

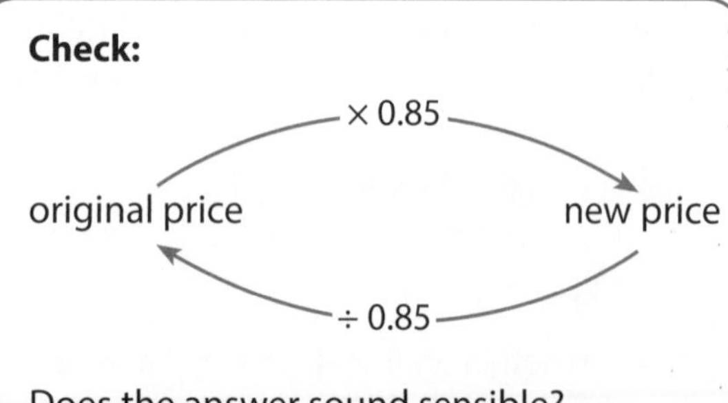

Does the answer sound sensible?

Is the original price more than the sale price?

Example 2

A mobile phone bill costs £169.20 including tax at 20%. What is the cost of the bill without the tax?

- The phone bill of £169.20 represents 100% + 20% = 120% of the original bill (x).
- 120% = 1.20 ← This is the **multiplier**.
- $1.2 \times x = £169.20$

$$x = \frac{£169.20}{1.2}$$

Original bill is **£141**

Check:

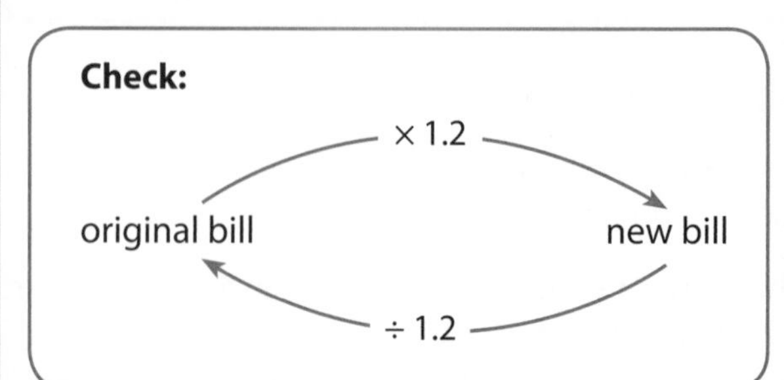

Example 3

The price of a washing machine is reduced by 5% in the sales. It now costs £323. What was the original price?

- The sale price is 100% – 5% = 95% of the pre-sale price (x).
- 95% = 0.95 ← This is the **multiplier**.
- $0.95 \times x = £323$

  $x = \frac{£323}{0.95}$

  Original price was **£340**

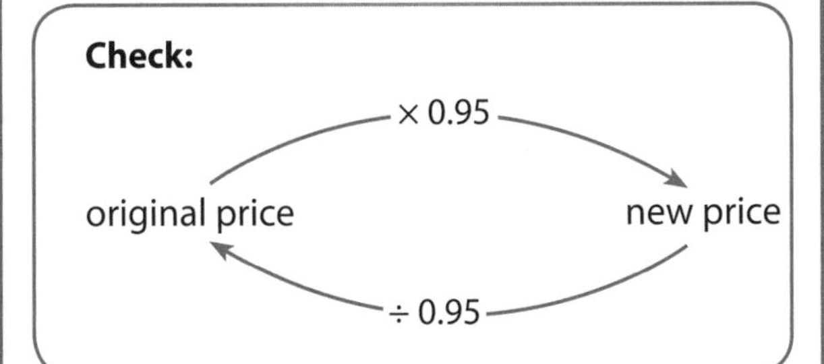

SUMMARY

- **A reverse percentage problem is where you know the amount after a percentage change and want to find the original amount.**
- **Use a multiplier to work out reverse percentage problems.**
- **Always check that your answer seems sensible.**

QUESTIONS

QUICK TEST

1. Each item listed below includes value added tax at 20%. Work out the original price of the item.

   a. A pair of shoes: £69

   b. A coat: £152.40

   c. A suit: £285

   d. A television: £525

EXAM PRACTICE

1. In a sale, normal prices are reduced by 12%. The sale price of a television is £220.

   Work out the normal price of the television.

   [3 marks]

2. Joseph says that the original price of a tablet device, which now costs £60 after a 15% reduction, was £70.59

   Is Joseph correct? Show your working.

   [3 marks]

Ratio and Proportion

Sharing a Quantity in a Given Ratio

A **ratio** is used to compare two or more related quantities. In **similar** shapes corresponding sides are in the same ratio.

The ratio 6 : 12 can be simplified to give 1 : 2

To share an amount in a given ratio, add up the individual parts and then divide the amount by this number to find one part.

Example
£155 is divided in the ratio of 2 : 3 between Daisy and Tom. How much does each receive?

$2 + 3 = 5$ parts ← Add up the total parts.

5 parts = £155

1 part = £155 ÷ 5 ← Work out what one part is worth.

= £31

So Daisy gets 2 × £31 = **£62**
and Tom gets 3 × £31 = **£93**

Check: £62 + £93
= £155 ✔

Exchange Rates

Two quantities are in **direct proportion** when both quantities increase at the same rate.

Example
Samuel went on holiday to Spain. He changed £350 into euros. The exchange rate was £1 = €1.36. How many euros did Samuel receive?

£1 = €1.36 so £350 = 350 × 1.36

= **€476**

Best Buys

Use unit amounts to help you decide which is the better value for money.

Example
The same brand of breakfast cereal is sold in two different-sized packets. Which packet represents better value for money?

Find the cost per gram for both boxes of cereal.

125 g costs £0.89 so $\frac{89}{125} = 0.712$p per gram

500 g costs £2.10 so $\frac{210}{500} = 0.42$p per gram

Since the larger box costs less per gram, it represents better value for money.

N.B. There are many different ways of working out the answer to this question.

Increasing and Decreasing in a Given Ratio

When increasing or decreasing in a given ratio, it is sometimes easier to find a unit amount.

Example

A recipe for four people needs 1600 g of flour. How much flour is needed to make the recipe for six people?

- Divide 1600 by 4, so 400 g for one person.
- Multiply by 6.

So 6×400 g = **2400 g** of flour is needed for six people.

When two quantities are in **inverse proportion**, one quantity increases at the same rate as the other quantity decreases. For example, the time it takes to build a wall increases as the number of builders decreases.

It took four builders six days to build a wall.

Time for four builders $= 6$ days

Time for one builder $= 6 \times 4 = 24$ days

It takes one builder four times as long to build the wall.

At the same rate it would take six builders $\frac{24}{6} = 4$ days

Maps and Diagrams

Scales are often used on maps and diagrams. They are usually written as **ratios**.

Example

The scale on a road map is 1 : 25 000. Watford and St Albans are 60 cm apart on the map. Work out the real distance between them in km.

On a scale of 1 : 25 000, 1 cm on the map represents 25 000 cm on the ground.

60 cm represents $60 \times 25\,000 = 1\,500\,000$ cm.

$1\,500\,000 \div 100 = 15\,000$ m ← Divide by 100 to change cm to m.

$15\,000 \div 1000 = 15$ km ← Divide by 1000 to change m to km.

The distance between Watford and St Albans is **15 km**.

SUMMARY

- **A ratio is used to compare two or more related quantities.**
- **Two quantities are in direct proportion when both quantities increase at the same rate.**
- **Two quantities are in inverse proportion when one quantity increases at the same rate as the other quantity decreases.**

QUESTIONS

QUICK TEST

1. Divide £160 in the ratio 1 : 2 : 5
2. The cost of four ringbinders is £6.72
   Work out the cost of 21 ringbinders.
3. It took six builders four days to lay a patio. At the same rate how long would it take eight builders?

EXAM PRACTICE

1. Toothpaste is sold in three different-sized tubes.

   50 ml is £1.24

   75 ml is £1.96

   100 ml is £2.42

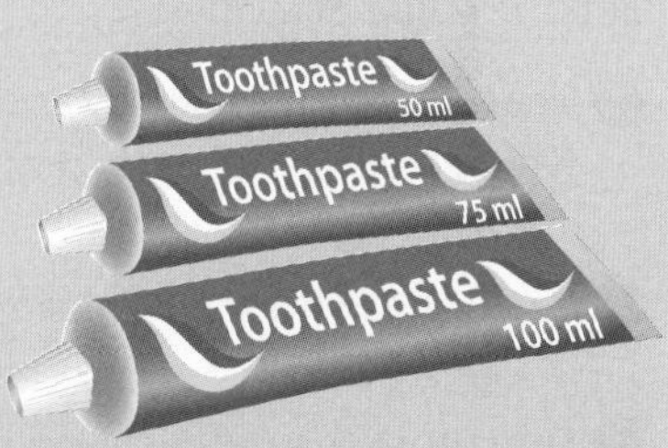

   Which of the tubes of toothpaste is the best value for money? You must show full working in order to justify your answer. [3 marks]

2. Jessica buys a pair of jeans in England for £52. She then goes on holiday to America and sees an identical pair of jeans for \$63. The exchange rate is £1 = \$1.49.

   In which country are the jeans cheaper, and by how much? [2 marks]

Proportionality

Direct Proportion

When one quantity increases in the same proportion as another quantity, the quantities are said to be in **direct proportion** to each other. For example, the cost of a bag of apples is directly proportional to the weight of the apples. The symbol $\propto$ is used to denote direct proportion.

If the cost of the bag of apples is C pence and the weight of the apples is W kg, then $C \propto W$. If the apples cost k pence per kilogram, $C = kW$

In general, if y is directly proportional to x:

$y \propto x$ and $y = kx$ where k is known as the **constant of proportionality.**

Since $y = kx$, the graph of y against x is a straight line passing through the origin.

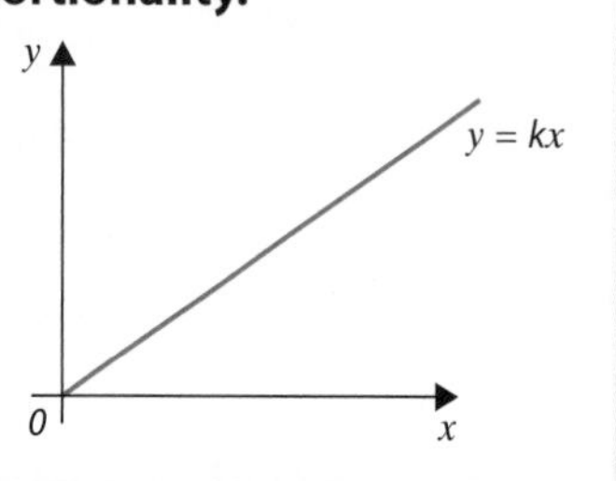

The constant of proportionality, k, is the gradient of this straight line.

Example 1

If a is proportional to b and $a = 5$ when $b = 4$, find the value of k (the constant of proportionality) and the value of a when $b = 8$.

- Change the sentence by adding the symbol $\propto$, which means 'is directly proportional to'.
  $a \propto b$
- Replace $\propto$ with '$= k$' to make an equation.
  $a = kb$
- Substitute the values given in the question in order to find k.
  $5 = k \times 4$
  $\frac{5}{4} = k$ ← Rearrange the equation.
- Replace k with the value just found.
  $a = \frac{5}{4}b$
  If $b = 8$ $\quad a = \frac{5}{4} \times 8$
  $a = \mathbf{10}$

Example 2

The voltage, V volts, across an electrical circuit is directly proportional to the current, I amps, flowing through the circuit.

When $I = 2.4$, $V = 156$

a. Work out the formula connecting V and I.

$V \propto I$

$V = kI$

$156 = k \times 2.4$ ← Substitute the values of $I = 2.4$ and $V = 156$ into the equation.

$\frac{156}{2.4} = k$

$k = 65$

$\boldsymbol{V = 65I}$

b. Find V when $I = 4$

$V = 65 \times 4$

$V =$ **260 volts**

c. Find I when $V = 357.5$ volts

$V = 65I$

$\frac{V}{65} = I$

$\frac{357.5}{65} = I$

$I =$ **5.5 amps**

Inverse Proportion

In **inverse proportion**, as one variable increases the other decreases, and as one variable decreases the other increases.

If y is inversely proportional to x, then you write this as

$y \propto \frac{1}{x}$ or $y = \frac{k}{x}$

The graph $y = \frac{k}{x}$, when k is positive, has a similar shape to $y = \frac{1}{x}$

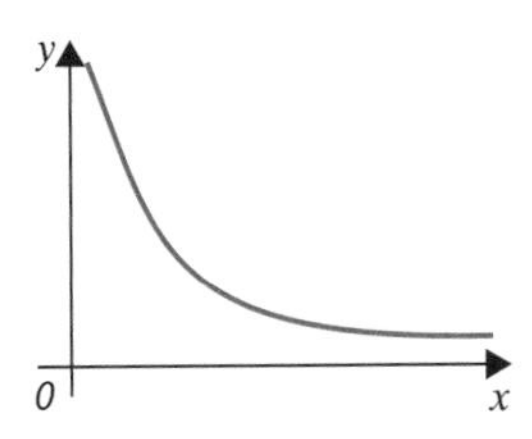

Example

p is inversely proportional to w. If $w = 2$ when $p = 5$, what is the value of w when $p = 20$?

$p \propto \frac{1}{w}$ ← Write the information with the proportionality sign.

$p = \frac{k}{w}$ ← Replace with the constant of proportionality.

$5 = \frac{k}{2}$

$5 \times 2 = k$

$k = 10$ ← Find the value of k.

$p = \frac{10}{w}$ ← Rewrite the formula.

$20 = \frac{10}{w}$ ← Find the value of w if $p = 20$.

$w = \frac{10}{20}$

$w = \mathbf{\frac{1}{2}}$ or $w = \mathbf{0.5}$

SUMMARY

- **The notation ∝ means 'is directly proportional to'. This is often abbreviated to 'is proportional to' or 'varies as'.**
- **In direct proportion, if one variable increases the other variable also increases and if one variable decreases, the other variable also decreases.**
- **In inverse proportion, if one variable increases the other variable decreases.**

QUESTIONS

QUICK TEST

1. Match these statements with the correct equation.

   a. y is proportional to x — $y = \frac{k}{x}$

   b. y is inversely proportional to x — $y = kx$

EXAM PRACTICE

1. a is directly proportional to x.
   When $x = 4$, $a = 8$.

   What is the constant of proportionality and what is the value of x when $a = 64$? [4 marks]

2. y is inversely proportional to x.
   When $x = 3$, $y = 10$.

   a. Calculate y when $x = 2$. [4 marks]

   b. Calculate x when $y = 6$. [2 marks]

Measurement

Units of Measurement

The **metric system** of units is based on tens. Older style units called **imperial units** are not, in general, based on tens.

Metric units		
Length	**Weight**	**Capacity**
10 mm = 1 cm	1000 mg = 1 g	1000 ml = 1 l
100 cm = 1 m	1000 g = 1 kg	100 cl = 1 l
1000 m = 1 km	1000 kg = 1 tonne	1000 cm^3 = 1 l

Comparisons between metric and imperial units		
Length	**Weight**	**Capacity**
2.5 cm ≈ 1 inch	28 g ≈ 1 ounce	1 litre ≈ $1\frac{3}{4}$ pints
8 km ≈ 5 miles	1 kg ≈ 2.2 pounds	4.5 litres ≈ 1 gallon
N.B. These comparisons are only approximate.		

Time Measurement

Time can be measured using the 12 or 24-hour clock.

The 12-hour clock uses am and pm. Times before midday are am, and times after midday are pm.

The 24-hour clock numbers the hours from 0 to 24. Times are written with four figures.

For example:

- 2.42 pm is the same as 1442
- 5.30 am is the same as 0530
- 1527 is the same as 3.27 pm
- 0704 is the same as 7.04 am

Units of Time

- There are 60 seconds in one minute.
- There are 60 minutes in one hour.
- There are 24 hours in one day.
- There are seven days in one week.
- There are 52 weeks in one year.
- There are 365 days in a year, or 366 in a leap year.

24-hour clock times often appear on bus and train timetables.

Conversions for Area and Volume

Changing Area Units

$1\ m^2 = 10\,000\ cm^2$

This square has a length of 1 metre. This is the same as a length of 100 cm.

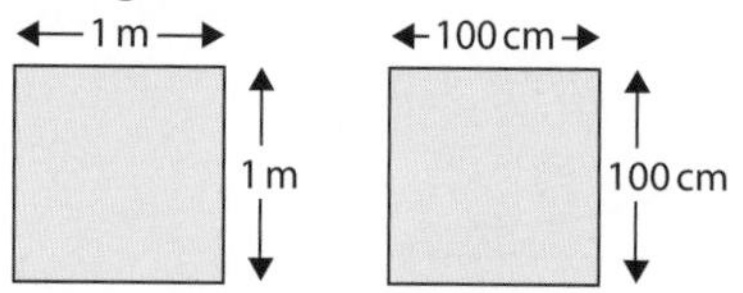

Hence $1\ m^2 = 100 \times 100\ cm^2$

$1\ m^2 = 10\,000\ cm^2$

Always check that the measurements are in the same units before you calculate an area.

Changing Volume Units

This cube has a length of 1 m. This is the same as a length of 100 cm.

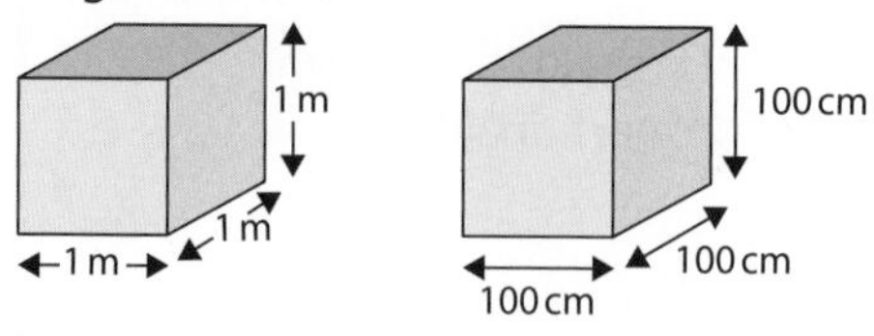

Hence $1\ m^3 = 100 \times 100 \times 100\ cm^3$

$1\ m^3 = 1\,000\,000\ cm^3$

Compound Measures

Compound measures involve a combination of basic measures.

Speed

Units of speed are:

- metres per second (m/s)
- kilometres per hour (km/h)
- miles per hour (mph)

Converting between compound measures should always be done in steps.

Example

Change 80 km/h into m/s.

	80 km	in 1 hour
is	80 000 m	in 1 hour
is	80 000 m	in 60 minutes
is	1333.333... m	in 1 minute
is	**22.22... m**	**in 1 second**

$$\text{Speed } (s) = \frac{\text{distance travelled } (d)}{\text{time taken } (t)}$$

Rearranging gives:

$$\text{Time taken} = \frac{\text{distance travelled}}{\text{speed}}$$

$$\text{Distance travelled} = \text{speed} \times \text{time taken}$$

Remember to check the units before starting a question. Change them if necessary.

Examples

1. A car travels 80 km in 1 hour 20 minutes. Find the speed in km/h.

$s = \frac{d}{t}$

$s = \frac{80}{1.\dot{3}}$ ← Change the time into hours. 20 minutes is $\frac{20}{60}$ of 1 hour.

$s =$ **60 km/h**

2. Miss Fitzgerald drives 40 miles to work. On Wednesday her journey to work took 50 minutes. On Thursday the average speed of her journey to work was 54 km/h.

Did Miss Fitzgerald drive more quickly to work on Wednesday or Thursday?

Speed on Wednesday is $\frac{40}{\frac{50}{60}} = 48$ mph

Since 1 km $\approx \frac{5}{8}$ mile

Speed on Thursday is $\frac{5}{8} \times 54 = 33.75$ mph

Miss Fitzgerald drove more quickly to work on Wednesday.

Density

$$\text{Density} = \frac{\text{mass}}{\text{volume}} \qquad \text{Volume} = \frac{\text{mass}}{\text{density}}$$

$$\text{Mass} = \text{volume} \times \text{density}$$

Example

Find the density of an object with a mass of 600 g and a volume of 50 cm^3.

Density $= \frac{M}{V} = \frac{600}{50} =$ **12 g/cm^3**

Pressure

$$\text{Pressure} = \frac{\text{force on surface}}{\text{surface area}} \qquad P = \frac{f}{a}$$

SUMMARY

- **Speed** $= \frac{\text{distance travelled}}{\text{time taken}}$
- **Density** $= \frac{\text{mass}}{\text{volume}}$
- **Pressure** $= \frac{\text{force on surface}}{\text{surface area}}$

QUESTIONS

QUICK TEST

1. Change 3200 g into kilograms.
2. What is 3.30 am written in 24-hour clock time?
3. Find the time taken for a car to travel 240 miles at an average speed of 70 mph.

EXAM PRACTICE

1. Two solids each have a volume of 2.5 m^3. The density of solid A is 320 kg per m^3. The density of solid B is 288 kg per m^3. Calculate the difference in the masses of the solids. [3 marks]
2. Josie is driving in her car at a speed of 70 mph. Jack is driving in his car at a speed of 120 km/h. Given that 5 miles ≈ 8 km, who is travelling at the fastest speed? You must show full working to explain your answer. [3 marks]

Interpreting Graphs

Conversion Graphs

Conversion graphs are used to change one unit of measurement into another unit; for example, litres to pints, kilometres to miles, pounds to dollars, etc.

Measurements such as kilometres and miles are in direct proportion to each other. A conversion graph of this kind of information can be drawn.

For example, if £1 = $1.50

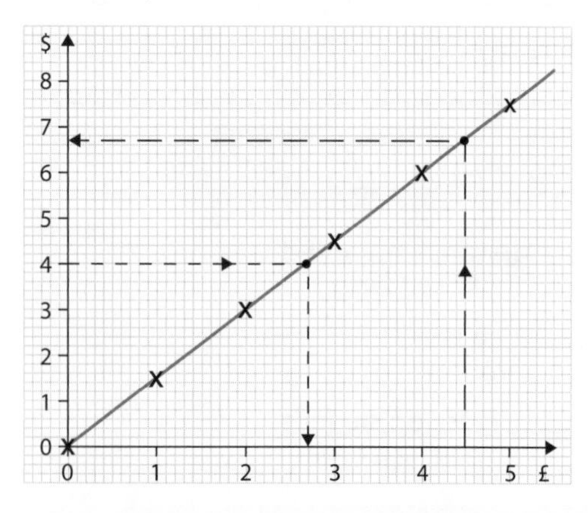

To change dollars to £, read across to the line and then read down: for example, $4 = £2.70 (approx.) To change £ to dollars, read up to the line then read across; for example, £4.50 is $6.80 (approx.)

Distance–Time Graphs

Distance–time graphs are often known as travel graphs. Distance is on the vertical axis; time is on the horizontal axis.

The **speed** of an object can be found on a distance–time graph by finding the **gradient** at a given time.

Example

Mr Smith travels from Leeds to his office 80 miles away.

The distance–time graph shows his journey.

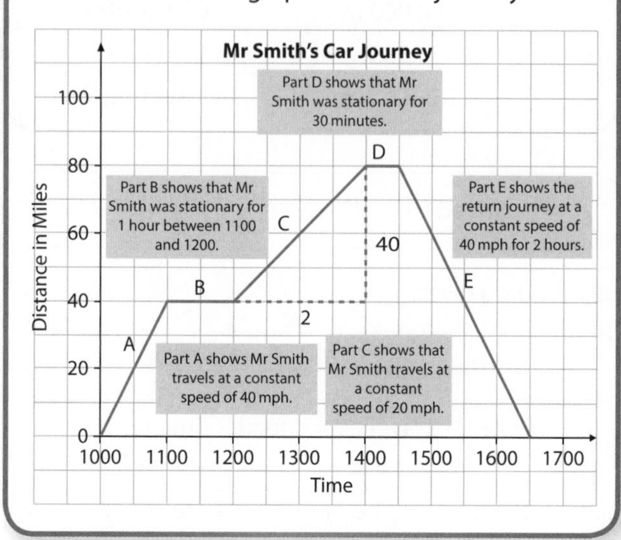

Always check that you understand the scales on a distance–time graph. For this graph:

- vertically one square represents 10 miles
- horizontally two squares represents 1 hour

The gradient of the distance–time graph represents the speed over that time interval.

$$\text{Speed} = \frac{\text{distance travelled}}{\text{time taken}}$$

In part C of the graph:

$$\text{speed} = \frac{40}{2}$$

speed = 20 mph

Speed–Time Graphs

A **speed–time** graph is useful when finding the **acceleration** or **deceleration** of an object. It can also be used to find the distance travelled.

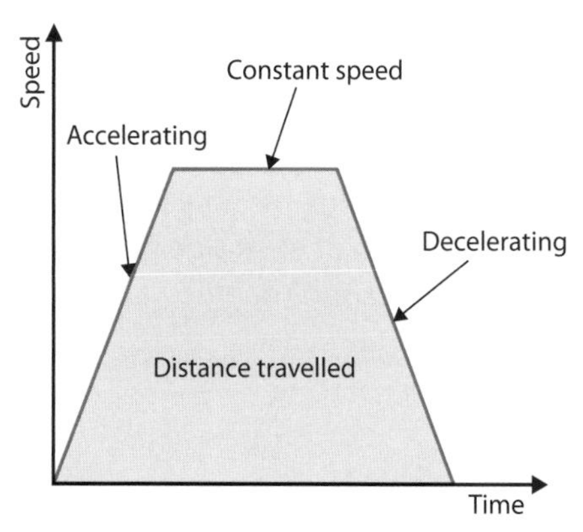

- Distance is the area between the graph and x-axis.
- A positive gradient means the speed is increasing.
- A negative gradient means the speed is decreasing.
- A horizontal line means the speed is constant.

In a speed–time graph, the gradient represents the acceleration at a given time.

SUMMARY

- **Remember to check the scales on a distance–time graph. The gradient on a distance–time graph represents the speed.**
- **Speed =** $\dfrac{\textbf{distance travelled}}{\textbf{time taken}}$
- **Speed–time graphs are useful when finding the acceleration or deceleration of an object. In a speed–time graph, the gradient represents the acceleration at a given time.**
- **The area under a speed–time graph represents the distance that the object has travelled.**

QUESTIONS

QUICK TEST

1. A car travels 120 kilometres in 2.5 hours. What is the speed of the car?
2. John walks 4.5 miles in 1.5 hours. Work out his speed.
3. Water is poured into these containers at a constant rate. Match each container to the correct graph.

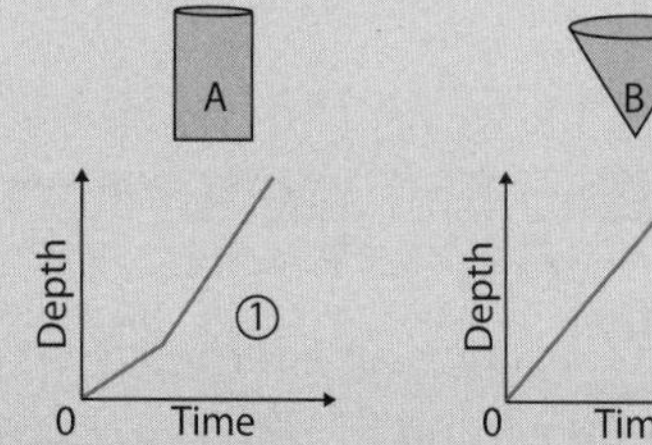

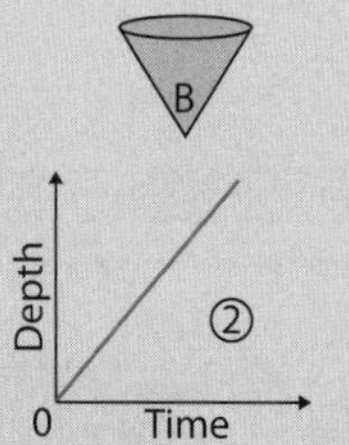

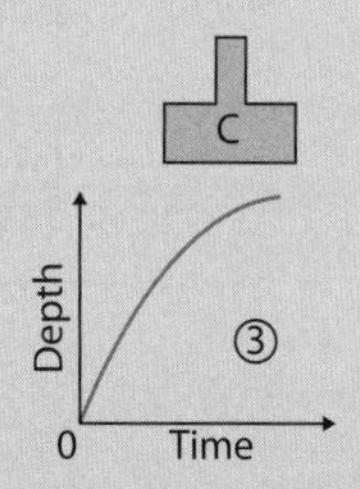

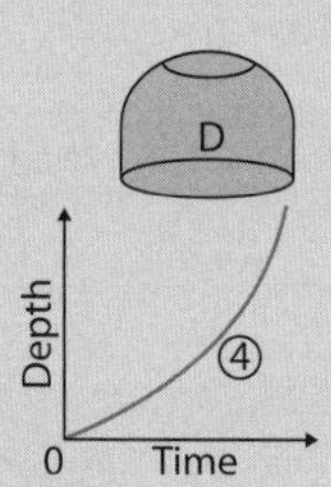

EXAM PRACTICE

1. The distance–time graph shows the car journeys of two people.

   Use the distance–time graph to work out the answers to these questions.

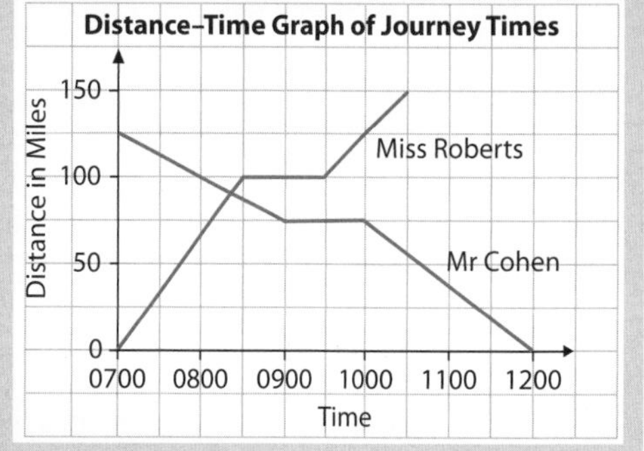

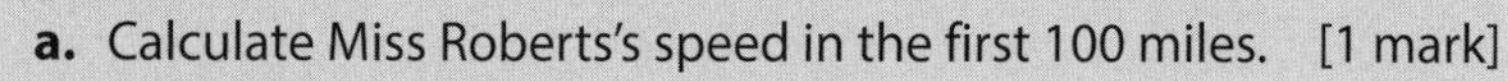

   a. Calculate Miss Roberts's speed in the first 100 miles. [1 mark]
   b. How long was Mr Cohen stationary for? [1 mark]
   c. Calculate Mr Cohen's speed between 1000 and 1200. [1 mark]

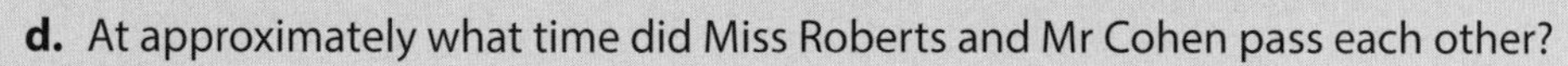

   d. At approximately what time did Miss Roberts and Mr Cohen pass each other? [1 mark]
   e. Calculate Miss Roberts's speed between 0930 and 1030. [1 mark]

Similarity

Similar Shapes

Objects that are exactly the same shape but different sizes are called **similar** shapes. One is an enlargement of the other.

Corresponding angles are equal.

Corresponding lengths are in the same ratio.

These shapes are similar. Each side of the first triangle is twice the length of each side of the second triangle.

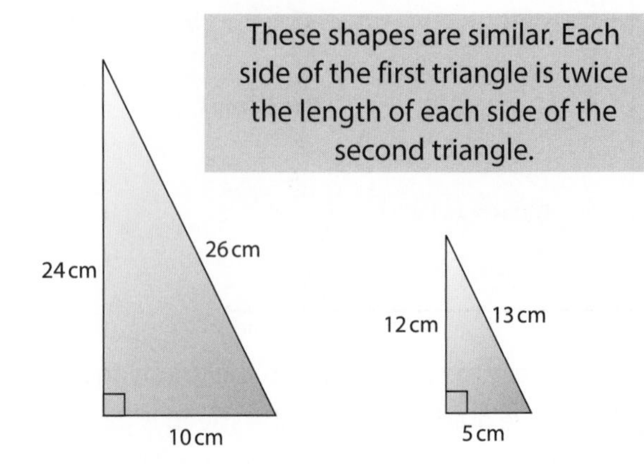

Finding Missing Lengths of Similar Figures

Questions about finding missing lengths are common at GCSE.

Examples

1. Find the missing lengths labelled a in the diagrams below:

a.

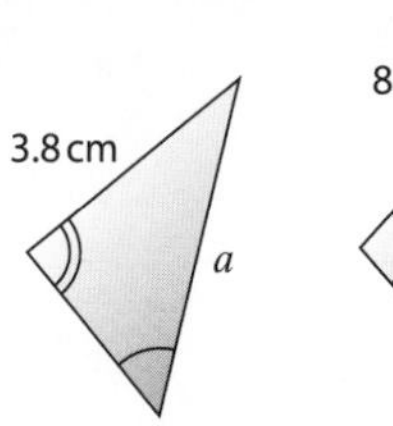

8.5 cm

12 cm

$\frac{a}{12} = \frac{3.8}{8.5}$ ← Corresponding lengths are in the same ratio.

$a = \frac{3.8}{8.5} \times 12$ ← Multiply both sides by 12.

$a =$ **5.36 cm** (3 s.f.)

b.

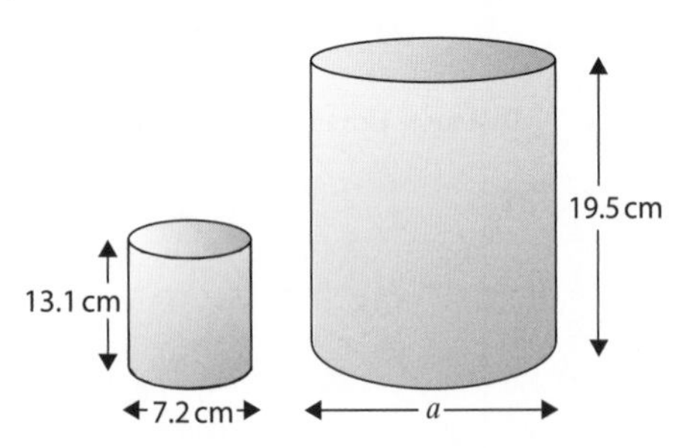

$\frac{a}{7.2} = \frac{19.5}{13.1}$

$a = \frac{19.5}{13.1} \times 7.2$ ← Multiply both sides by 7.2

$a =$ **10.7 cm** (3 s.f.)

2. Calculate the missing length y.

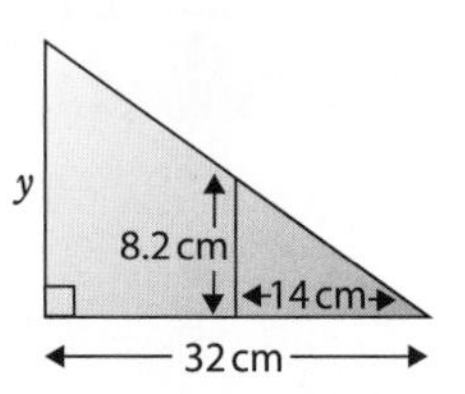

First draw the individual triangles.

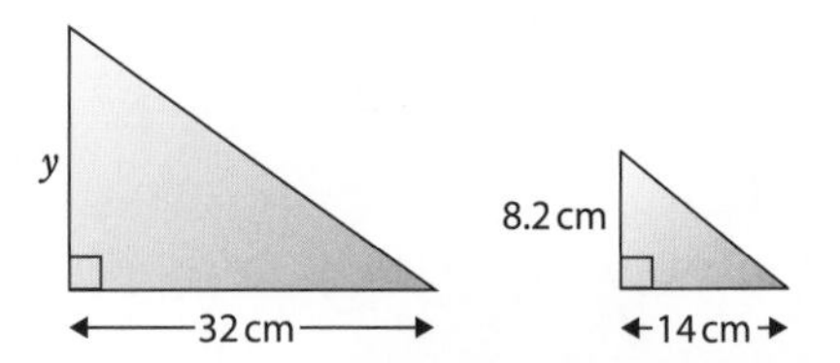

$\frac{y}{32} = \frac{8.2}{14}$ ← This gives an alternative way to writing the ratios as seen in the other examples. Both are correct!

$y = \frac{8.2}{14} \times 32$ ← Multiply both sides by 32.

$y =$ **18.7 cm** (3 s.f.)

More Difficult Problems

Sometimes you will be given a more difficult problem to solve.

Example

In the diagram CD is parallel to EF.

$EF = 4.1$ cm $FG = 5$ cm $DG = 7.2$ cm $CG = 12$ cm

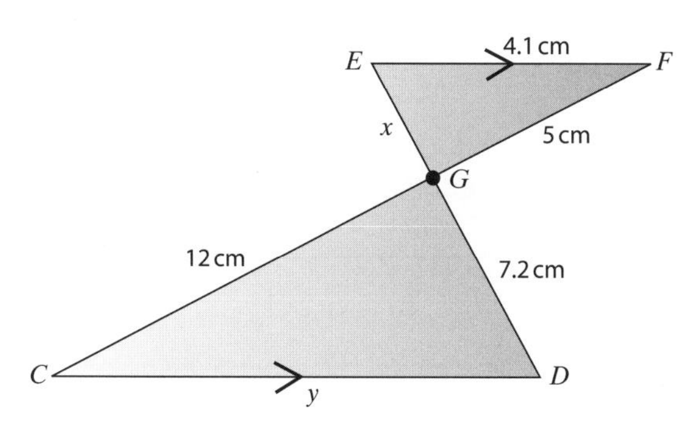

a. Explain why triangles EFG and CDG are similar.

The triangles are similar because:

- Angle FEG equals angle GDC (alternate angles are equal)
- Angle EFG equals angle GCD (alternate angles are equal)
- Angle EGF equals angle CGD (vertically opposite angles are equal)

The triangles are similar, since all three angles are the same.

b. Calculate the lengths marked x and y.

To find x:

$$\frac{x}{5} = \frac{7.2}{12}$$

$$x = \frac{7.2}{12} \times 5$$

$x =$ **3 cm**

To find y:

$$\frac{y}{12} = \frac{4.1}{5}$$

$$y = \frac{4.1}{5} \times 12$$

$y =$ **9.84 cm**

SUMMARY

- **Objects that are exactly the same shape but different sizes are called similar shapes.**
- **In similar shapes, corresponding angles are equal.**
- **In similar shapes, corresponding lengths are in the same ratio.**

QUESTIONS

QUICK TEST

1. Calculate the lengths marked n in these similar shapes. Give your answers correct to 1 d.p.

a.

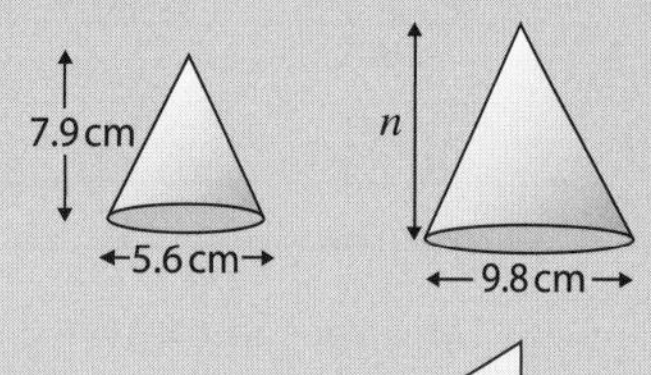

b.

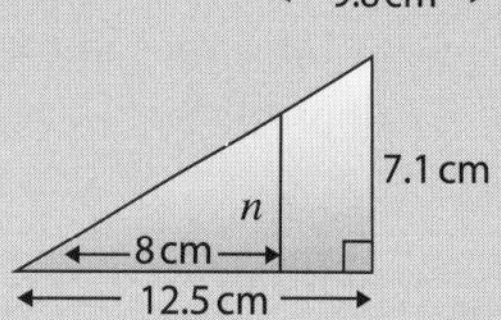

EXAM PRACTICE

1. Lucy says that these two triangles are similar. Is Lucy correct? Give a reason for your answer. [2 marks]

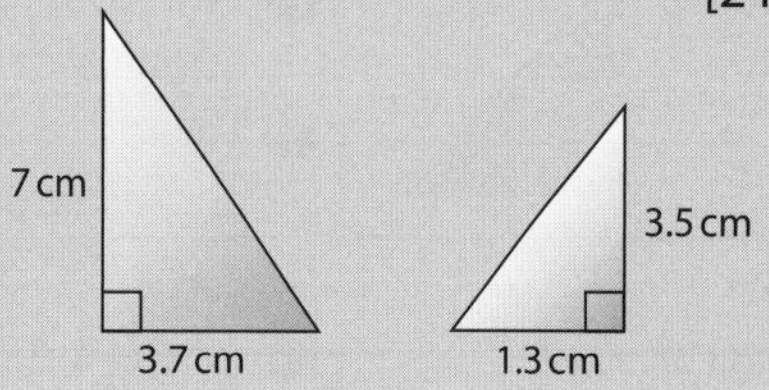

2. These two triangles are similar. Work out the missing length x. [3 marks]

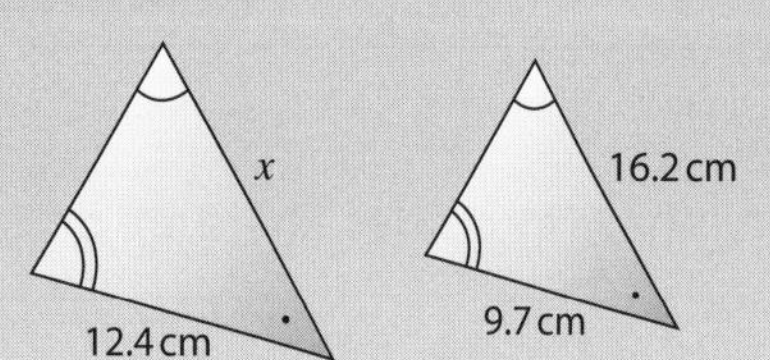

2D and 3D Shapes

2D Shapes

2D (two-dimensional) shapes have area. All points on 2D shapes are in the same plane.

Polygons

Polygons are 2D shapes with straight sides. Regular polygons are shapes with all sides and angles equal.

Number of sides	Name of polygon
3	Triangle
4	Quadrilateral
5	Pentagon
6	Hexagon
7	Heptagon
8	Octagon
9	Nonagon
10	Decagon

Regular pentagon
- Five equal sides
- Rotational symmetry of order 5
- Five lines of symmetry

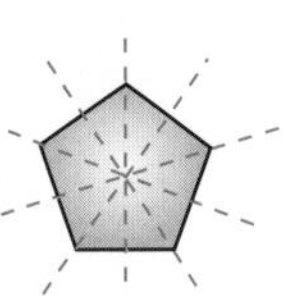

Triangles

Triangles have three sides. There are different types:

Right-angled
- Has a 90° angle.

Equilateral
- Three sides equal.
- Three angles equal.

60° 60° 60°

Isosceles
- Two sides equal.
- Base angles equal.

Scalene
- All the sides and angles are different.

Quadrilaterals

Quadrilaterals have four sides. There are different types:

Square
- Four lines of symmetry
- Rotational symmetry of order 4
- All angles are 90°
- All sides equal
- Two pairs of parallel sides
- The diagonals are equal and **bisect** each other at right angles.

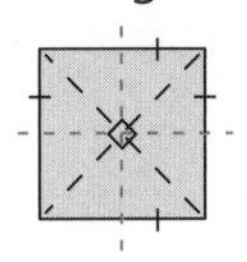

Rectangle
- Two lines of symmetry
- Rotational symmetry of order 2
- All angles are 90°
- Opposite sides equal
- Two pairs of parallel sides
- The diagonals bisect each other

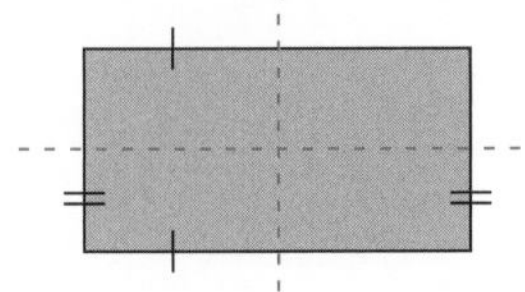

Parallelogram
- No lines of symmetry
- Rotational symmetry of order 2
- Opposite sides are equal and parallel
- Opposite angles are equal

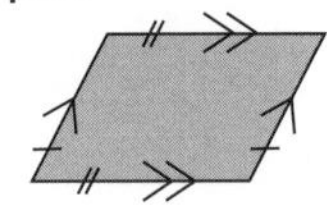

Rhombus
- Two lines of symmetry
- Rotational symmetry of order 2
- All sides are equal
- Opposite sides are parallel
- Opposite angles are equal
- The diagonals bisect each other at right angles and also bisect the corner angles

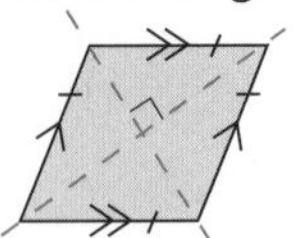

Kite
- One line of symmetry
- No rotational symmetry
- Diagonals do not bisect each other
- Two pairs of adjacent sides are equal
- Diagonals cross at right angles

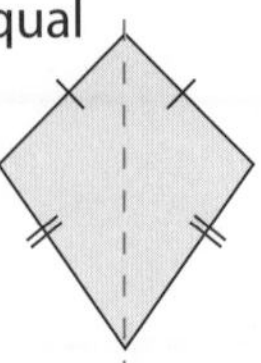

Trapezium
- Has one pair of parallel sides
- No lines of symmetry
- No rotational symmetry
- An isosceles trapezium has one line of symmetry

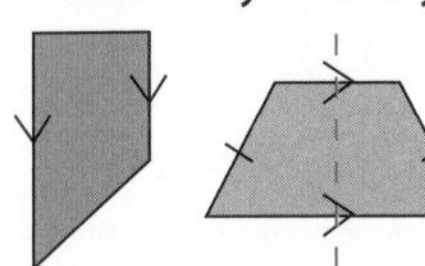

3D Solids

3D (three-dimensional) objects have volume (or capacity). You should know these 3D solids:

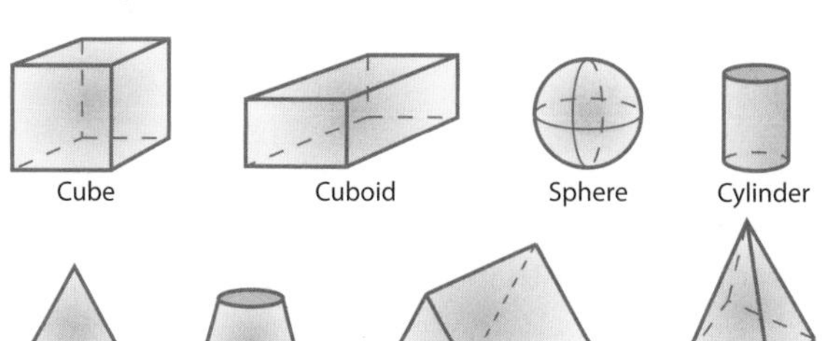

A **prism** is a solid which can be cut into slices that are all the same shape. Make sure you also know:

A **face** is a flat surface of a solid.	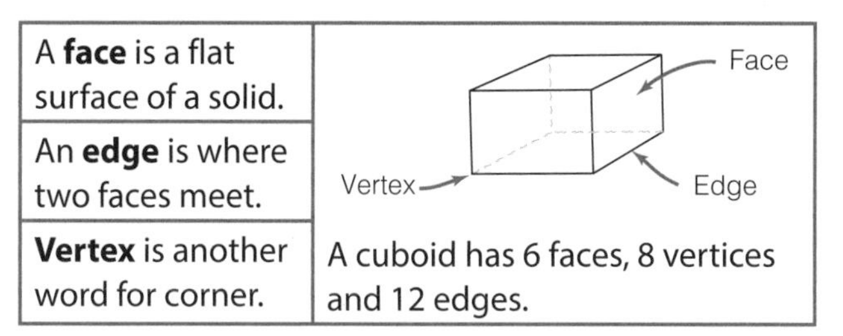
An **edge** is where two faces meet.	
Vertex is another word for corner.	A cuboid has 6 faces, 8 vertices and 12 edges.

The **net** of a 3D solid is a 2D (flat) shape that can be folded to make the 3D solid, e.g. for a cuboid:

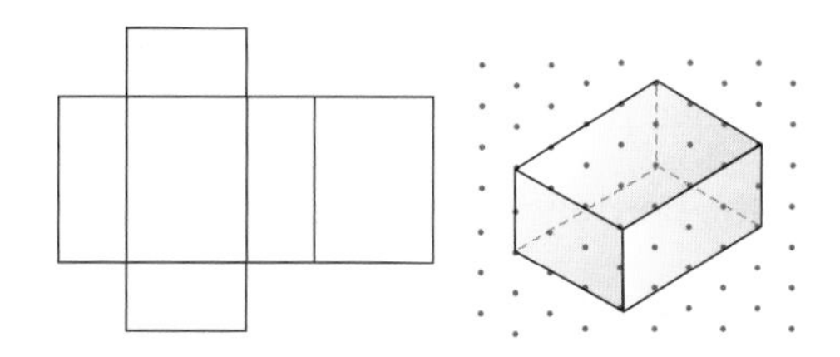

Plans and Elevations

- A **plan** is a view of a 3D solid from above.
- An **elevation** is seen if the 3D solid is looked at from the side or front.

Example

Draw a sketch of the plan and the elevations from A and B of this solid.

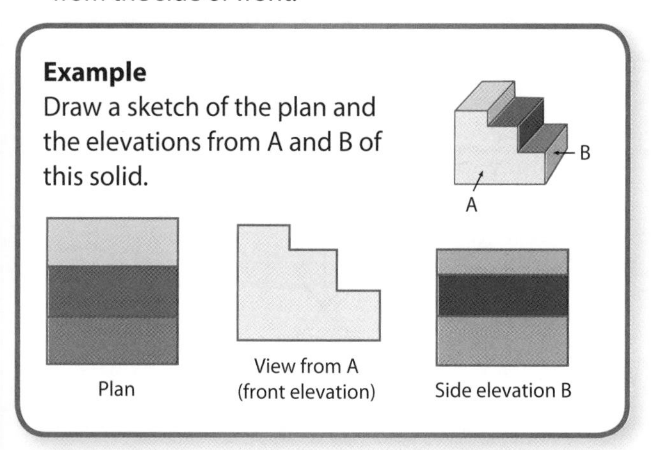

Plan

View from A (front elevation)

Side elevation B

SUMMARY

- **A plan is what can be seen if a 3D solid is looked down on from above.**
- **An elevation is seen if the 3D solid is looked at from the side or front.**
- **The net of a 3D solid is a 2D (flat) shape that can be folded to make the 3D solid.**

QUESTIONS

QUICK TEST

1. What is the name of a six-sided polygon?
2. Joe says: 'This triangle is right-angled'.

   Decide whether Joe is correct, giving a reason for your answer.

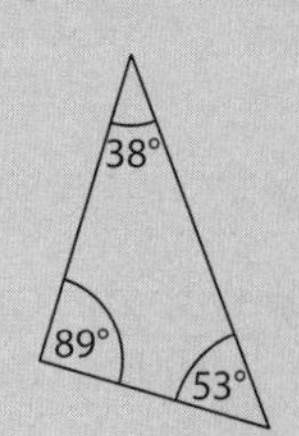

EXAM PRACTICE

1. Draw an accurate net of this 3D shape. [2 marks]

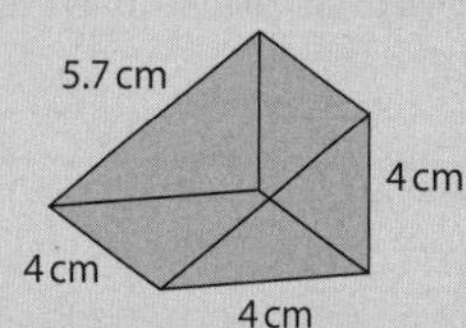

2. Here is a square-based pyramid.

   a. How many faces does it have? [1 mark]

   b. How many edges does it have? [1 mark]

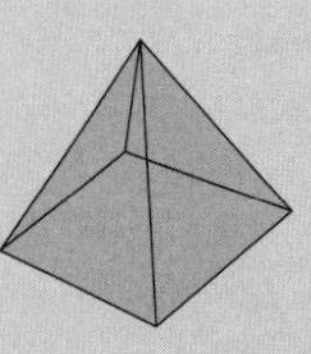

3. Draw the plan of this solid. [2 marks]

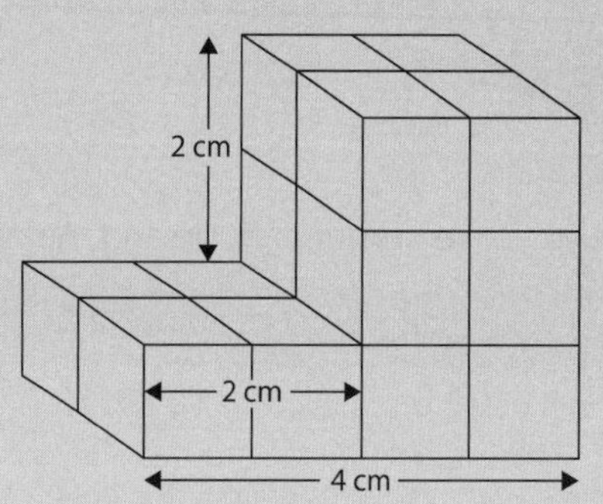

Constructions

The following constructions can be completed using only a ruler and a pair of compasses.

Constructing a Triangle

Use compasses to accurately construct this triangle.

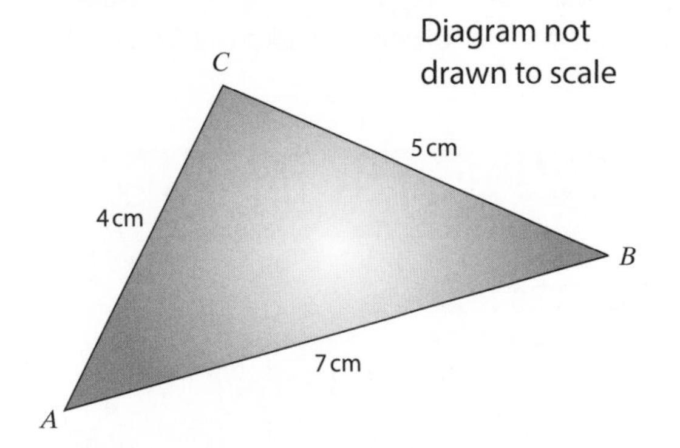

- Draw the longest side AB.
- With the compass point at A, draw an arc of radius 4 cm.
- With the compass point at B, draw an arc of radius 5 cm.
- Join A and B to the point where the two arcs meet at C.

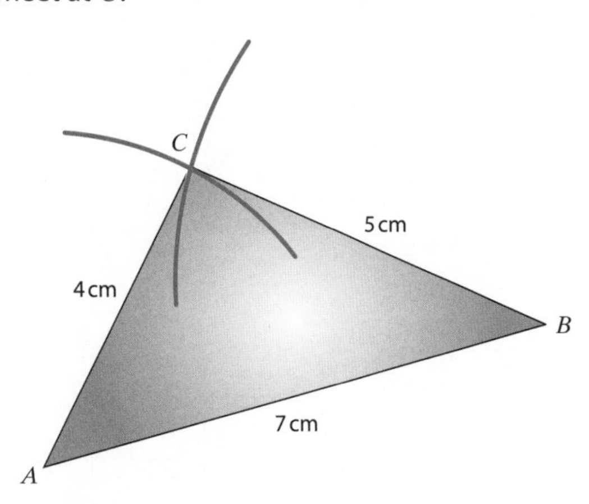

Remember to leave in your arcs. No arcs, no marks!

The Perpendicular Bisector of a Line

- Draw a line XY.
- Draw two arcs with the pair of compasses, using X as the centre. The pair of compasses must be set at a radius greater than half the distance of XY.

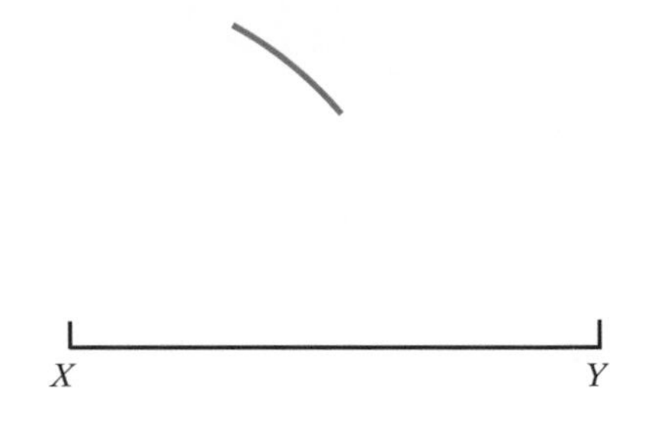

- Draw two more arcs with Y as the centre. (Keep the pair of compasses the same distance apart as before.)
- Join the two points where the arcs cross.
- AB is the **perpendicular bisector** of XY.
- N is the **midpoint** of XY.
- The perpendicular distance from a point to a line is the shortest distance to the line.

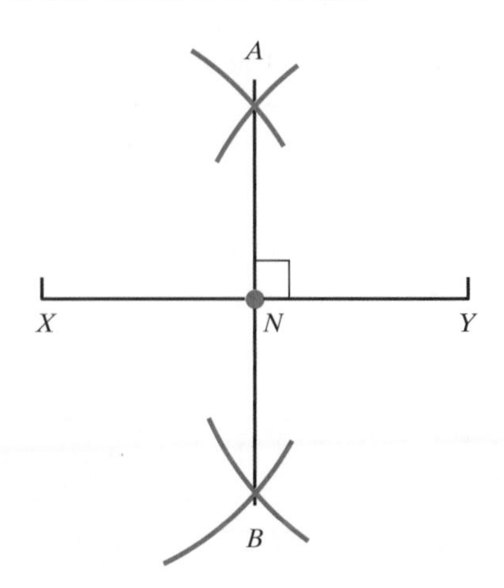

The Perpendicular from a Point to a Line

- From P draw arcs to cut the line at A and B.
- From A and B draw arcs with the same radius to intersect at C.
- Join P to C; this line is perpendicular to AB.

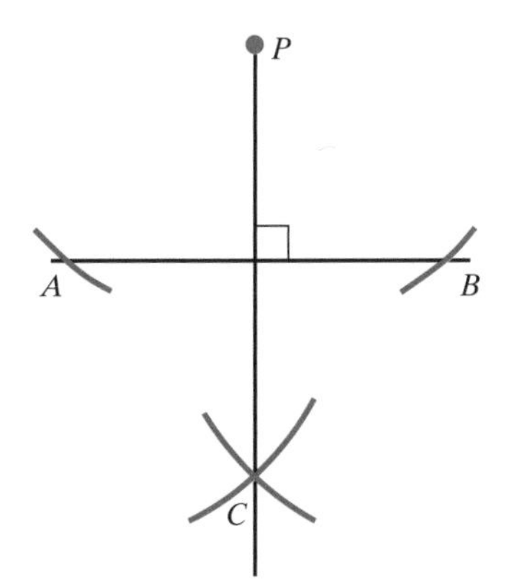

Bisecting an Angle

- Draw two lines XY and YZ to meet at an angle.

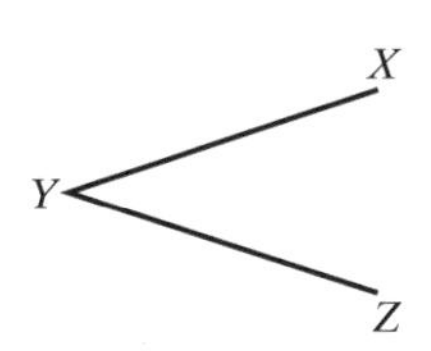

- Using a pair of compasses, place the point at Y and draw arcs on XY and YZ.

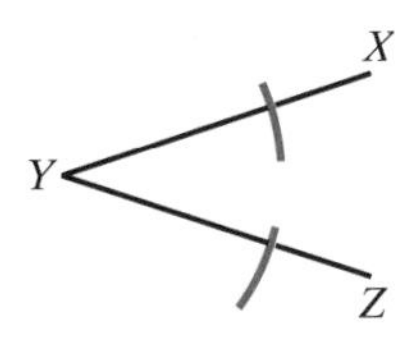

- Place the compass point at the two arcs on XY and YZ and draw arcs to cross at N. Join Y and N. YN is the bisector of angle XYZ.

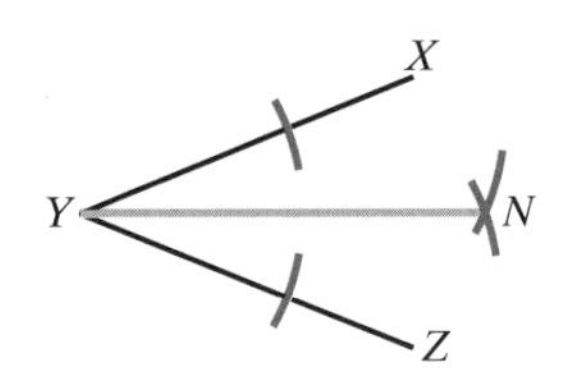

SUMMARY

- **Always leave in your construction arcs.**
- **Perpendicular means at right angles to.**
- **Bisect means to cut in half.**

QUESTIONS

QUICK TEST

1. Construct an angle of 30°, using only a ruler and a pair of compasses.
2. Draw the perpendicular bisector of an 8 cm line.

EXAM PRACTICE

1. Using only a ruler and a pair of compasses, construct the perpendicular from the point P.

   You must show all construction lines. [2 marks]

   • P

   A B

2. Using only a ruler and a pair of compasses, bisect this angle.

   You must show all construction lines. [2 marks]

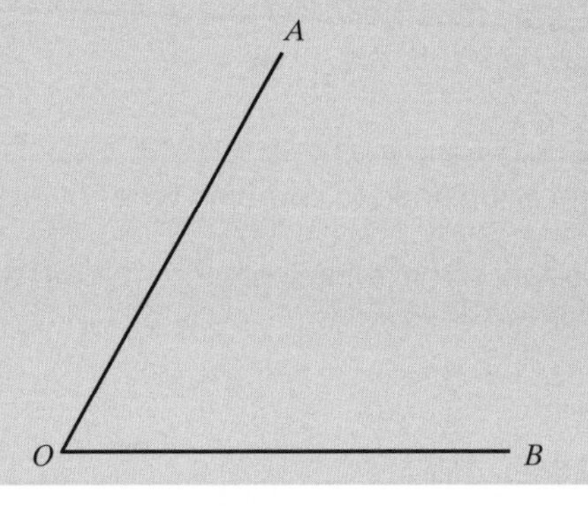

Loci

The **locus** of a point is the set of all the possible positions that the point can occupy subject to some given condition or rule.

Types of Loci

1. The locus of the points that are a constant distance from a fixed point is a circle.	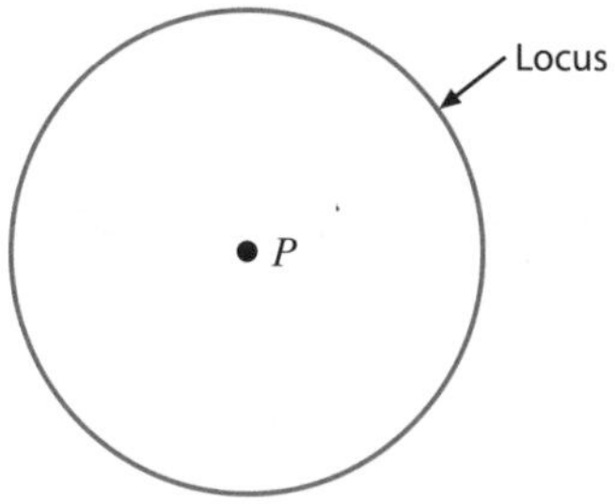
2. The locus of the points that are equidistant from two points X and Y is the perpendicular bisector of XY. *Two lines are perpendicular when they meet at 90°.* *The perpendicular distance from a point to a line is the shortest distance to the line.*	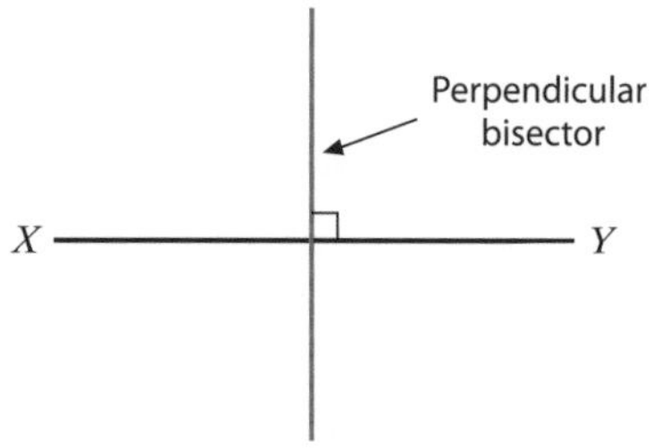
3. The locus of the points that are equidistant from two lines is the line that bisects the angle between the lines.	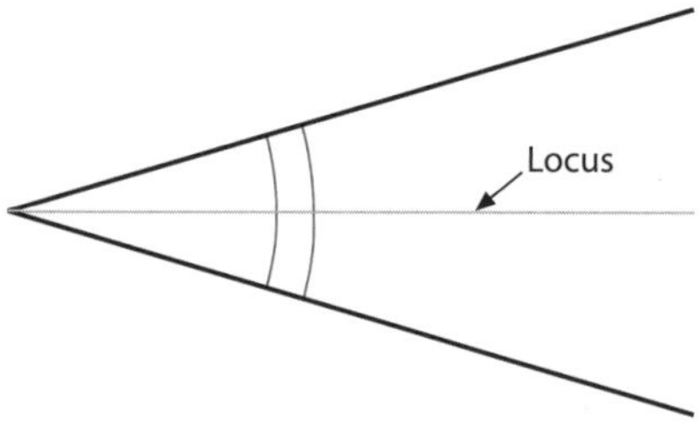
4. The locus of the points that are a constant distance from a line is a pair of lines parallel to the given line, one either side of it.	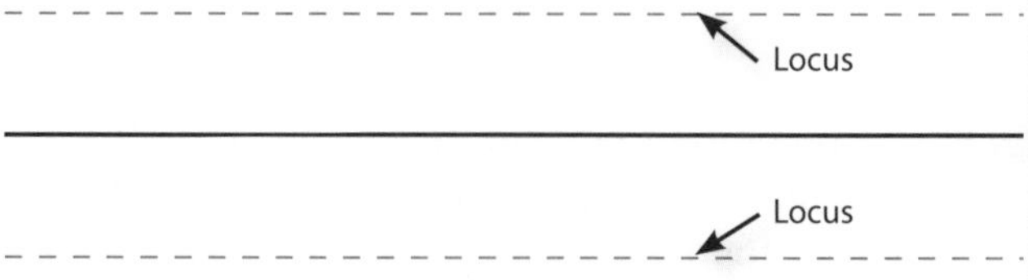
Sometimes you need to combine types of loci. A fixed distance from a line segment XY gives this locus.	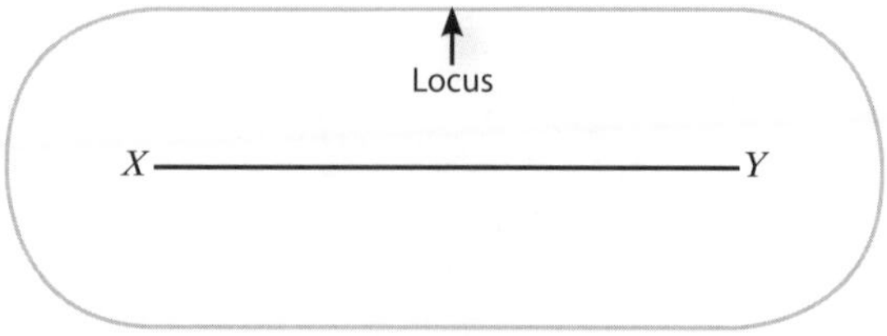

Example

Three radio transmitters form an equilateral triangle ABC with sides of 50 km. The range of the transmitter at A is 37.5 km, at B 30 km and at C 28 km. Using a scale of 1 cm to 10 km, construct a scale diagram to show where signals from all three transmitters can be received.

Below is a sketch not drawn to scale.

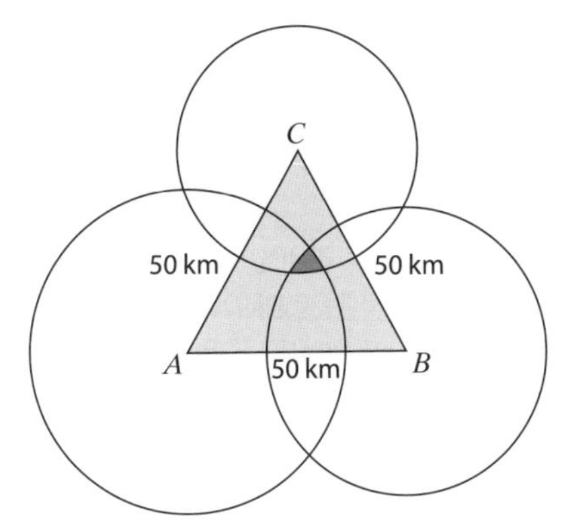

Please note that on your scale drawing the circle at A would have a radius of 3.75 cm. The circle at B would have a radius of 3 cm and the circle at C a radius of 2.8 cm.

The area where signals from all three transmitters can be received is shaded dark blue.

SUMMARY

- **Remember the four main types of loci:**
  - **A constant distance from a fixed point is a circle.**
  - **Equidistant from two fixed points is the perpendicular bisector of the line segment joining the points.**
  - **Equidistant from two lines is the bisector of the angle between the lines.**
  - **A constant distance from a line is a pair of parallel lines above and below.**

QUESTIONS

QUICK TEST

1. $ABCD$ is a rectangle. The rectangle is accurately drawn.

Shade the set of points inside the rectangle which are more than 2 cm from point B and more than 1.5 cm from the line AD.

EXAM PRACTICE

1. The plan shows a garden drawn to a scale of 1 cm : 2 m. A and B are bushes and C is a pond.

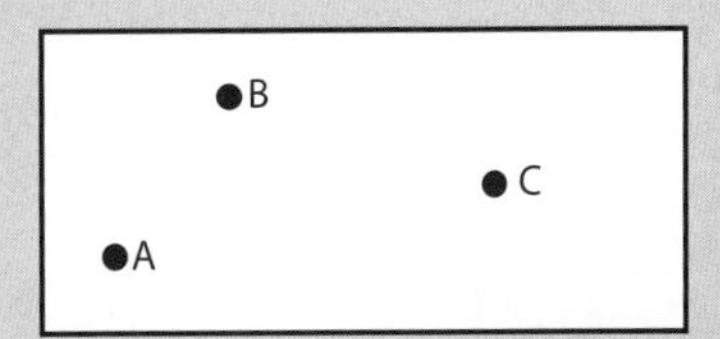

A landscape gardener has decided:

- to lay a path right across the garden at an equal distance from each of the bushes, A and B.
- to lay a flower border around pond C at a distance of 2 m.

Construct these features on the plan above.

[2 marks]

Angles

Types of Angle

- An **acute** angle is between 0° and 90°.
- An **obtuse** angle is between 90° and 180°.
- A **reflex** angle is between 180° and 360°.
- A **right angle** is 90°.

Acute angle

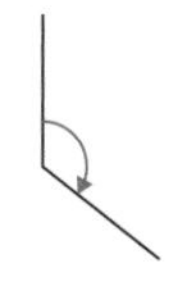
Obtuse angle

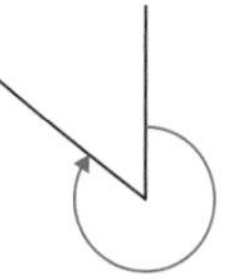
Reflex angle

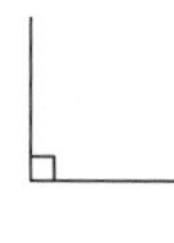
Right angle

Reading Angles

When asked to find angle ABC or $\angle ABC$ or $A\hat{B}C$, find the middle letter angle, i.e. at B:

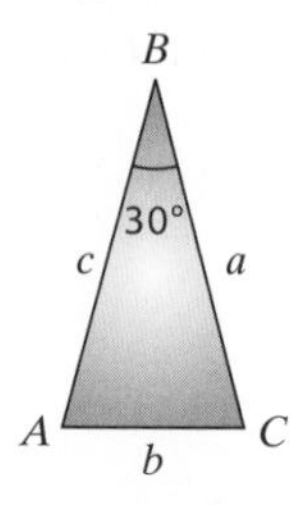

$\angle ABC = 30°$

When labelling a general triangle, the side opposite the vertex A is called a, the side opposite the vertex B is called b and the side opposite the vertex C is called c.

Angle Facts

Whenever lines meet or intersect, the angles they make follow certain rules:

Angles on a straight line add up to 180°. $a + b + c = 180°$ 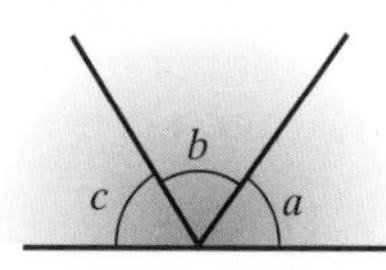	Angles in a triangle add up to 180°. $a + b + c = 180°$ 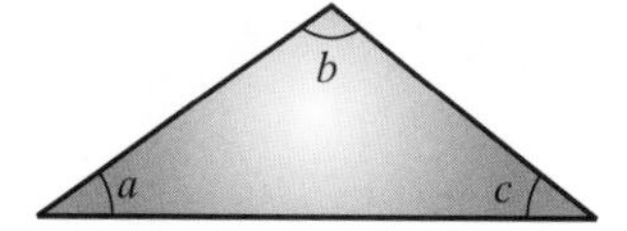	Angles in a quadrilateral add up to 360°. $a + b + c + d = 360°$ 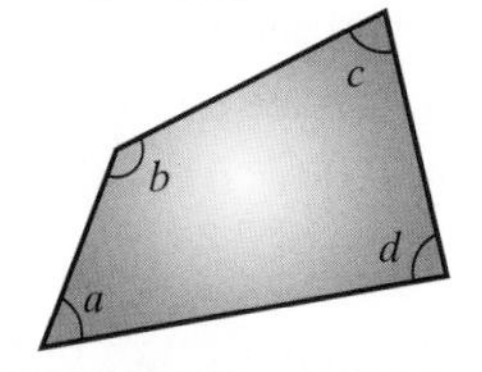
Angles at a point add up to 360°. $a + b + c + d = 360°$ 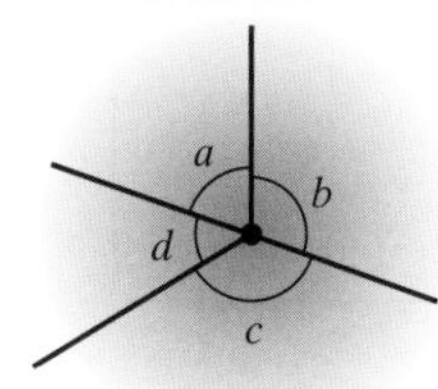	Vertically opposite angles are equal. $a = b, c = d$ $a + d = b + c = 180°$ 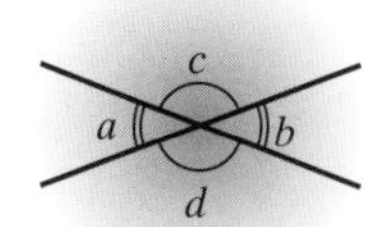	An exterior angle of a triangle equals the sum of the two opposite interior angles. $c = a + b$ 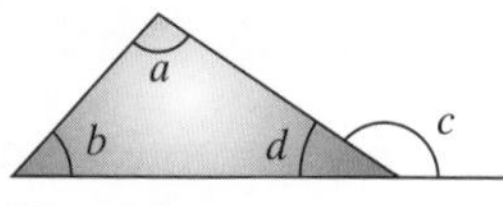 Since if $a + b + d = 180°$ (angles in a triangle add up to 180°) $d + c = 180°$ (angles on a straight line add up to 180°) Then $c = a + b$

Example

Find the missing angles in the diagram below.

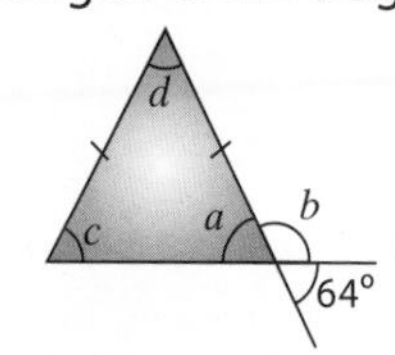

$\boldsymbol{a} = \mathbf{64°}$ (opposite angles are equal)

$\boldsymbol{b} = 180° - 64° = \mathbf{116°}$ (angles on a straight line add up to 180°)

$\boldsymbol{c} = \mathbf{64°}$ (isosceles triangle; base angles are equal)

$\boldsymbol{d} = \mathbf{52°}$ (angles in a triangle add up to 180°)

Parallel Lines

Three types of relationship are produced when a line called a transversal crosses a pair of parallel lines.

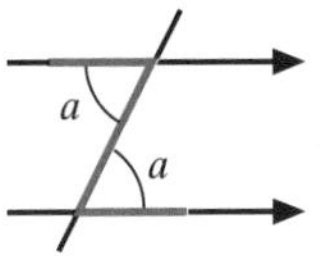

Alternate angles are equal.

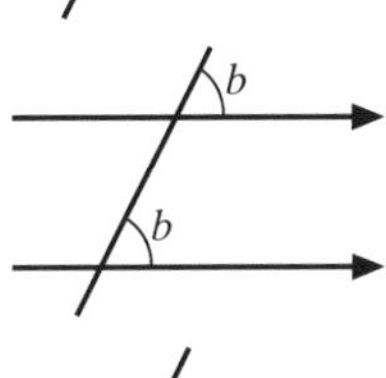

Corresponding angles are equal.

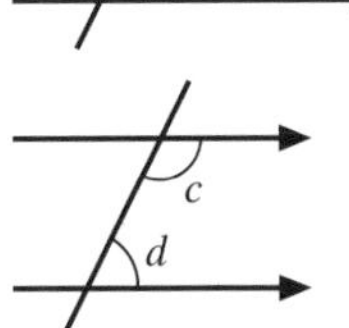

Allied or **supplementary** angles add up to 180°. $c + d = 180°$

Angles in Polygons

There are two types of angle in a polygon – **interior** and **exterior**. A regular polygon has all sides and angles equal.

For a polygon with n sides:

- Sum of exterior angles = 360°
- Interior angle + exterior angle = 180°
- Sum of interior angles = $(n - 2) \times 180°$ or $(2n - 4) \times 90°$

For a regular polygon with n sides:

- Exterior angle = $\frac{360°}{n}$

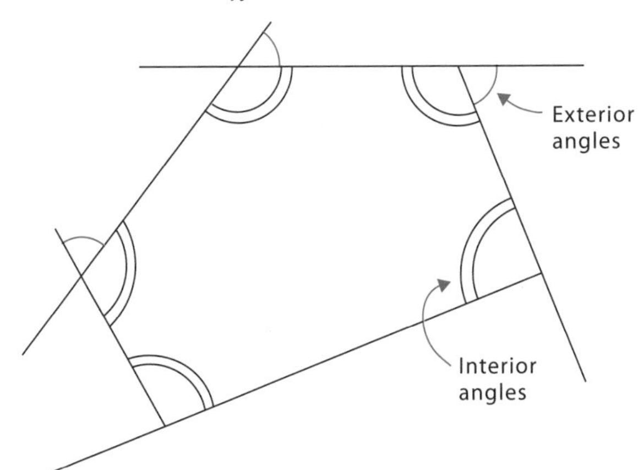

Example
A regular polygon has an interior angle of 108°. How many sides does it have?

180° – 108° = 72° (size of exterior angle)

360° ÷ 72° = **5 sides**

SUMMARY

- **Angles add up to: 180° on a straight line, 360° at a point, 180° in a triangle, 360° in a quadrilateral.**
- **Alternate and corresponding angles are equal; supplementary angles add up to 180°.**
- **Polygons have interior and exterior angles.**

QUESTIONS

QUICK TEST

1. Work out the sizes of the angles in the diagrams below.

a. **b.** **c.**

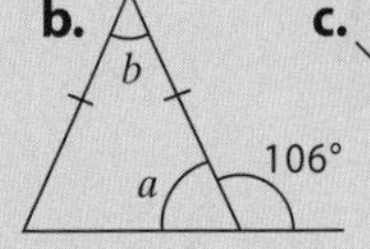

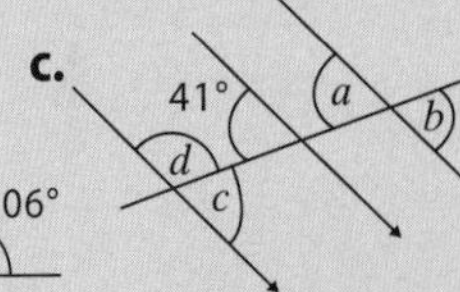

2. Work out the size of the exterior angle of a 12-sided regular polygon.

EXAM PRACTICE

1. BCD is a straight line. Explain why BE and CF must be parallel. [2 marks]

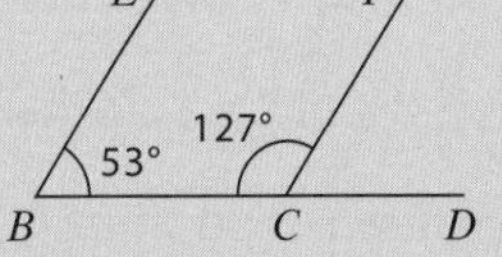

2. Work out the size of angle y in this polygon. [4 marks]

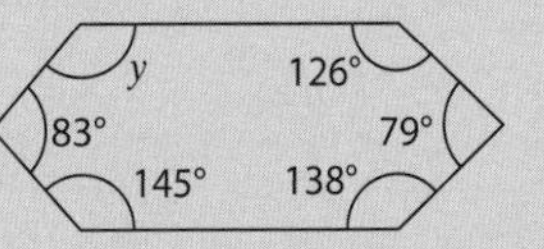

3. The diagram shows a regular octagon and a regular hexagon.

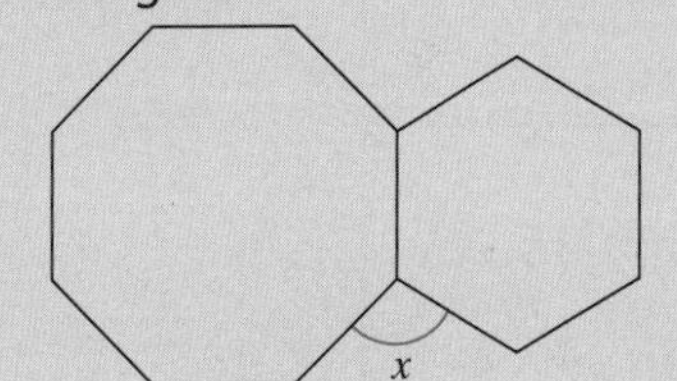

Find the size of the angle marked x. You must show all your working. [3 marks]

Bearings

Compass Directions

The diagram shows the points of the compass. Directions can also be given as **bearings**. Bearings are used on aeroplanes and ships to make sure they are travelling in the right direction and to avoid collisions.

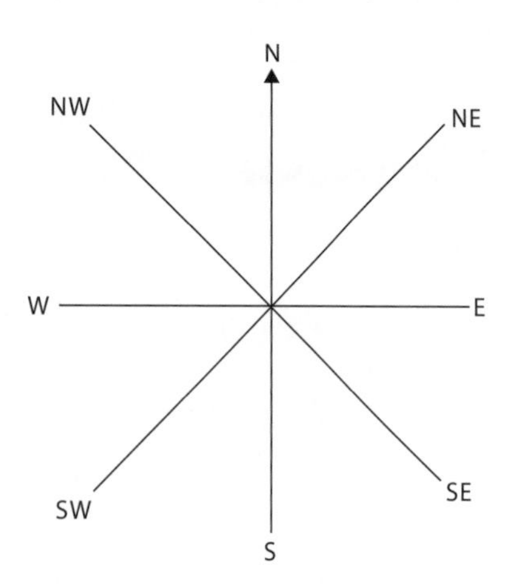

Bearings

Bearings are often used in scale drawing questions.

- They are always measured from the North (N).
- They are measured in a **clockwise** direction.
- They are written using three figures.

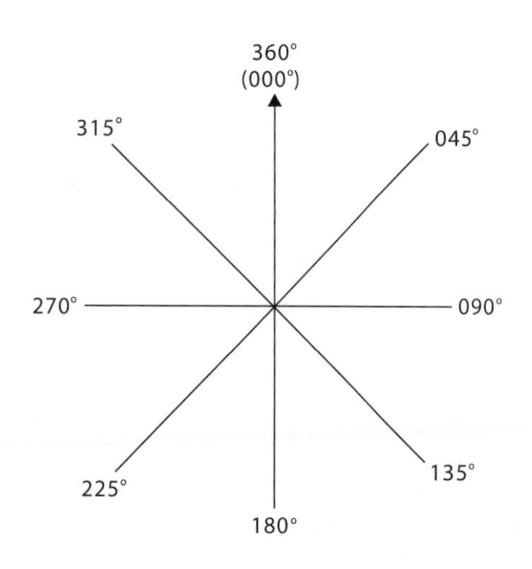

Examples

Find the bearing of P **from** Q in each diagram.

a.

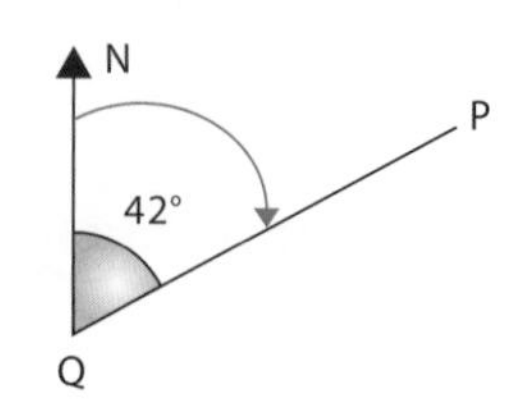

Bearing of P from Q:

= **042°**

Bearings must have three figures, so write 042°.

b.

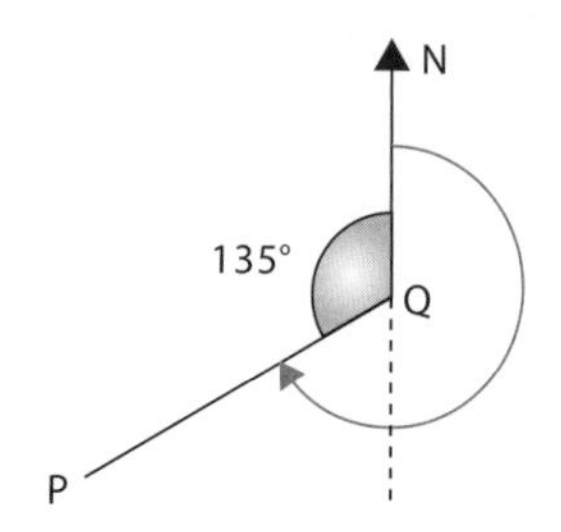

Bearing of P from Q:

= 360° – 135°

= **225°**

c.

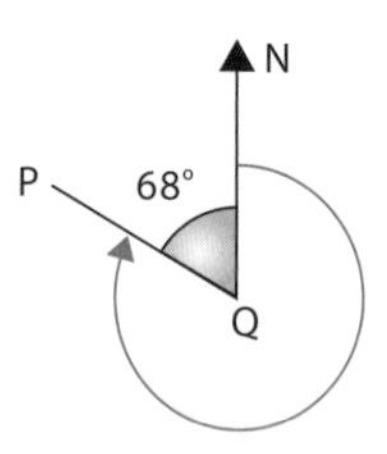

Bearing of P from Q:

= 360° – 68°

= **292°**

Back Bearings

Back bearings are more difficult – you need to draw in a second North line. The angle properties of parallel lines can then be used.

Examples
Find the bearing of Q from P in each diagram.

a.

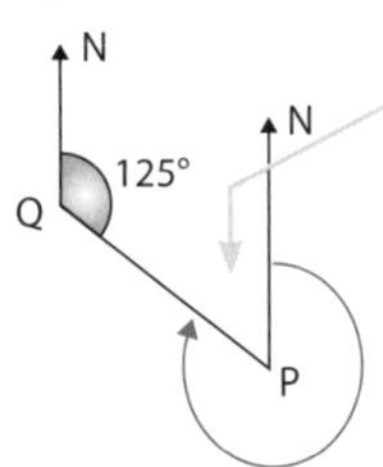

The North lines are parallel. Supplementary angles add up to 180°. 180° – 125° = 55°

Bearing of Q from P:
= 360° – 55°
= **305°**

b.

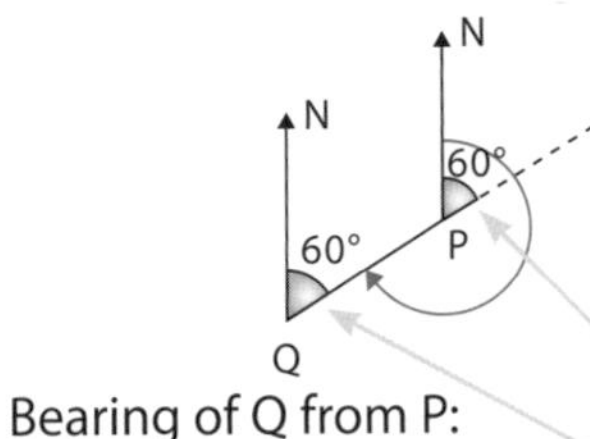

These angles are corresponding so are equal.

Bearing of Q from P:
= 60° + 180°
= **240°**

Scale Drawings and Bearings

Scale drawings are very useful for finding lengths that cannot be measured directly. When drawing scale diagrams, the lengths need to be accurate to 2 mm and the angles to 2°.

Example
A ship sails from a harbour for 15 km on a bearing of 040°, and then continues due east for 20 km. Make a scale drawing of this journey using a scale of 1 cm to 5 km. How far will the ship have to sail to get back to the harbour by the shortest route? What will the bearing be?

Shortest route = 6.4 × 5 km = **32 km**

Bearing = 70° + 180° = **250°**

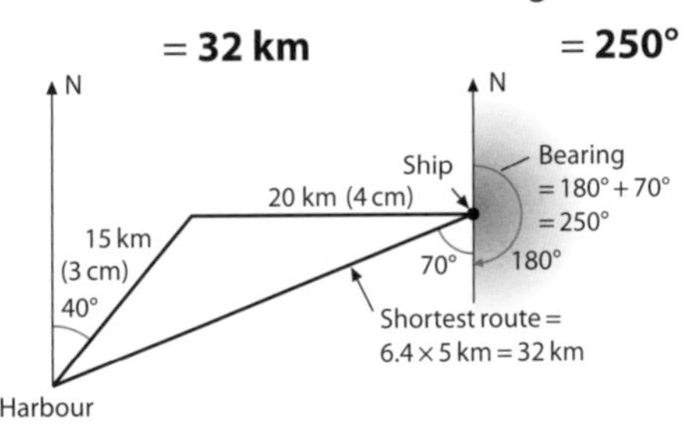

This diagram is not drawn to scale but is used to show what your diagram should look like.

SUMMARY

- **Remember the key facts about bearings:**
  - **They are always measured from the North.**
  - **They are measured in a clockwise direction.**
  - **They are written using three figures, for example 060°.**

QUESTIONS

QUICK TEST

1. For the following diagrams find the bearing of B from A.

a.

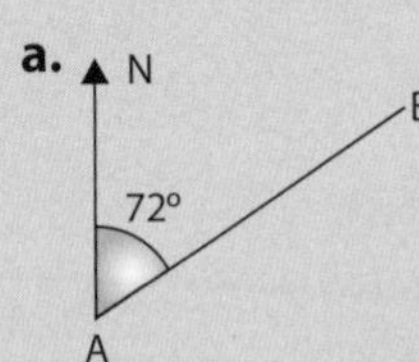

b.

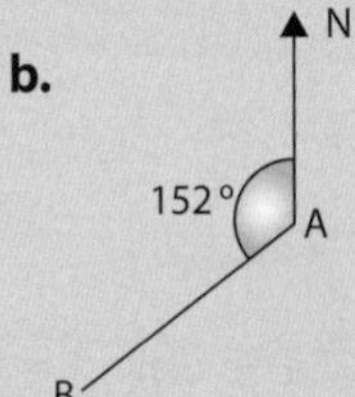

c.

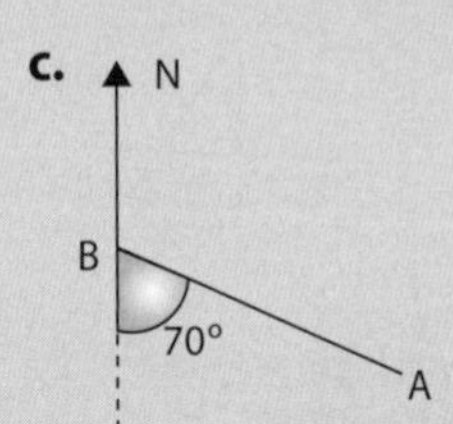

d.

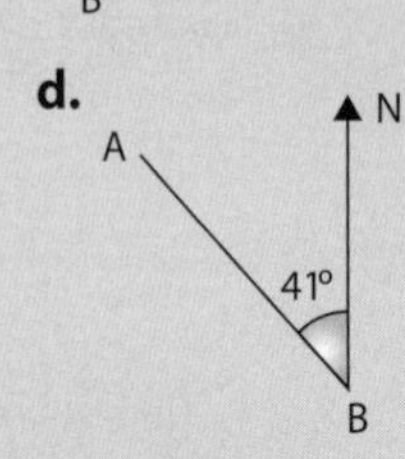

EXAM PRACTICE

1. The diagram shows the position of three towns A, B and C. Find the bearing of:

   a. A from B [1 mark]

   b. C from B [2 marks]

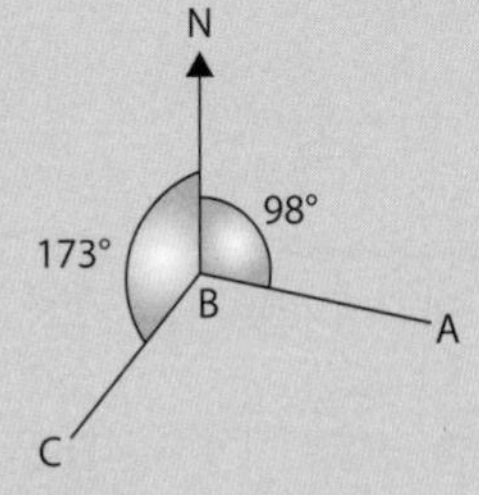

2. A ship sails on a bearing of 143° to a buoy, A.

   Work out the bearing the ship needs to sail to return to its starting point from the buoy, A.
   [3 marks]

Translations and Reflections

Translations

Translations move figures from one position to another position. **Vectors** are used to describe the distance and direction of the translations.

A vector is written as $\binom{a}{b}$.

a represents the horizontal distance and b represents the vertical distance.

The object and the image are **congruent** when the shape is translated, i.e. they are identical.

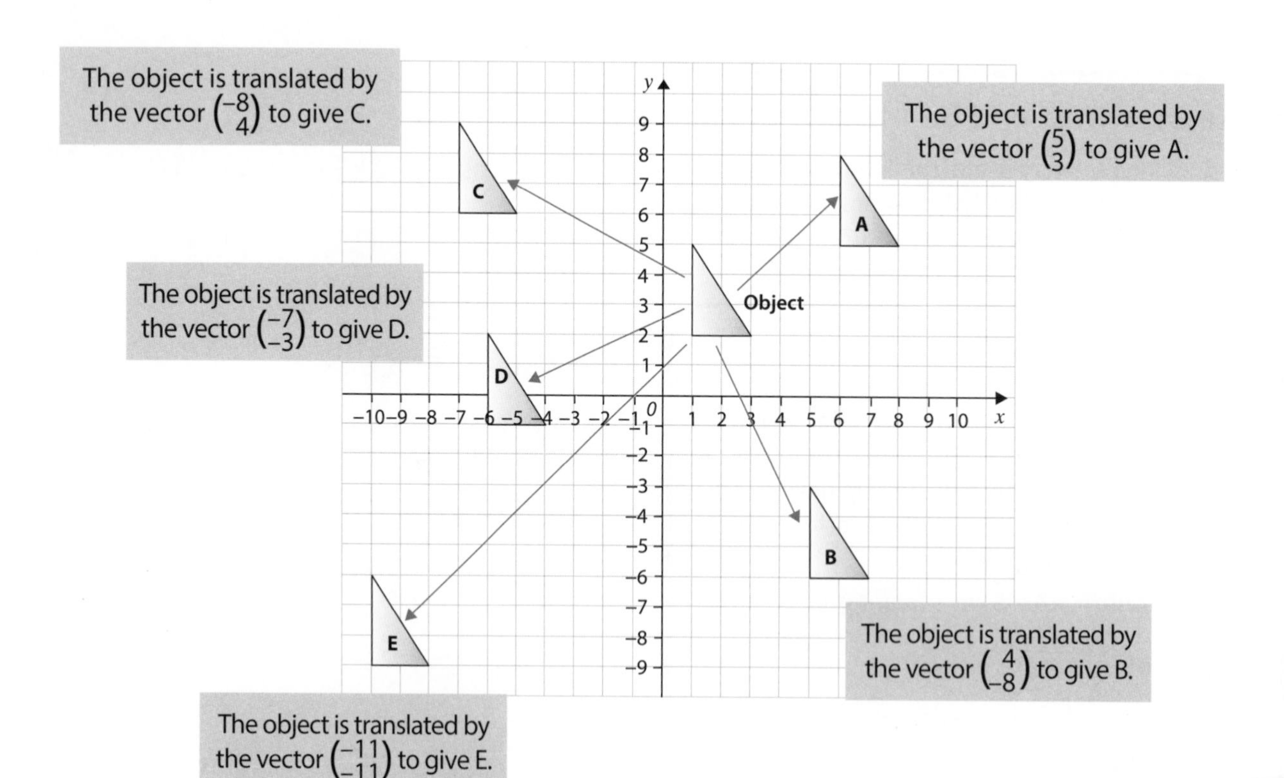

Reflections

Reflections create an image of an object on the other side of the mirror line.

The mirror line is known as an axis of reflection.

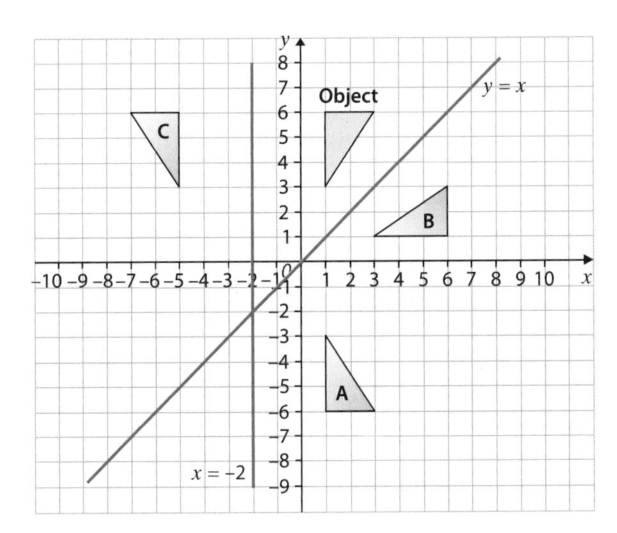

On the example above, the object is reflected in the x-axis (or $y = 0$) to give image A.

The object is reflected in the line $y = x$ to give image B.

The object is reflected in the line $x = -2$ to give image C.

The images and object are congruent.

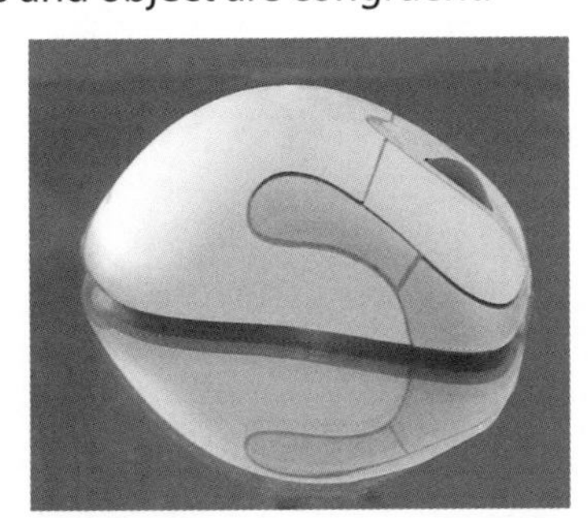

SUMMARY

- **Vectors are used to describe the distance and direction of translations.**
- **Reflections create an image of an object on the other side of the mirror line.**
- **The object and image are congruent when the shape is translated and reflected.**

QUESTIONS

QUICK TEST

1. For the diagram below, describe fully the transformation that maps:

   a. A onto B **b.** B onto C

   c. A onto D **d.** A onto E

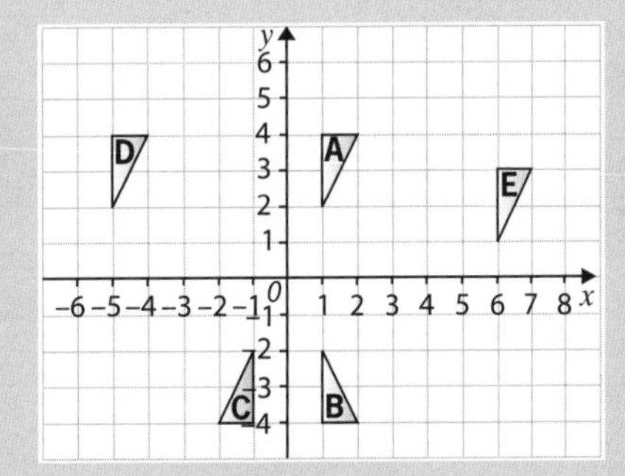

EXAM PRACTICE

1. Triangles A and B are shown on the grid below.

   a. Reflect triangle A in the line $x = 5$
   Label this image C. [2 marks]

   b. Translate triangle B by the vector $\begin{pmatrix} 4 \\ 2 \end{pmatrix}$
   Label this image D. [1 mark]

   c. Describe fully the single transformation which will map triangle A onto triangle B. [2 marks]

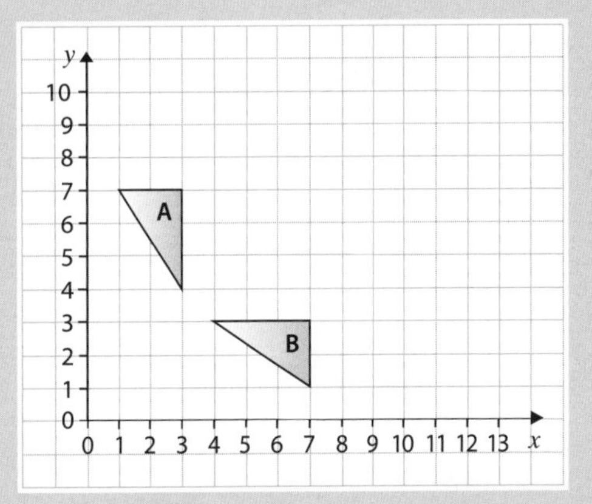

Rotation, Enlargement and Congruency

Rotations

In a **rotation** the object is turned by a given angle about a fixed point called the **centre of rotation**. The size and shape of the figure are not changed, i.e. the image is **congruent** to the object.

On the example below, object A is rotated by 90° clockwise about (0, 0) to give image B.

Object A is rotated by 180° about (0, 0) to give image C.

Object A is rotated 90° anticlockwise about (–2, 2) to give image D.

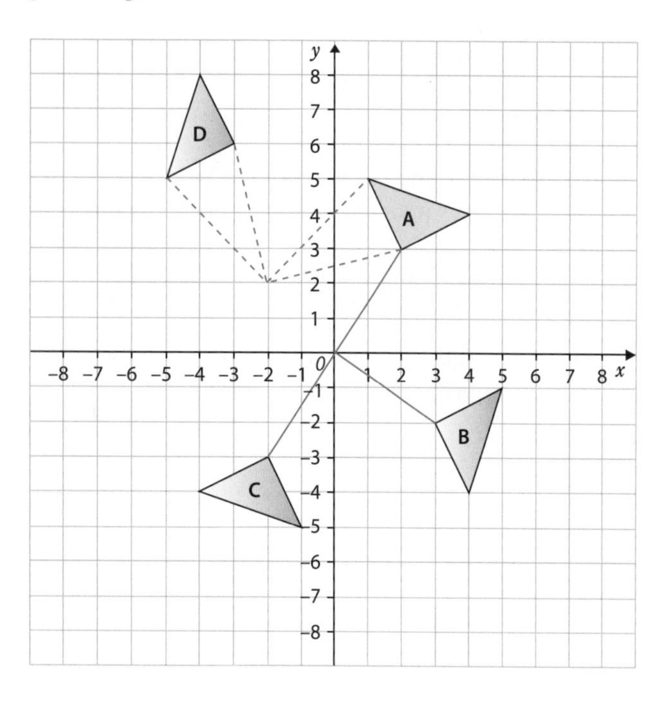

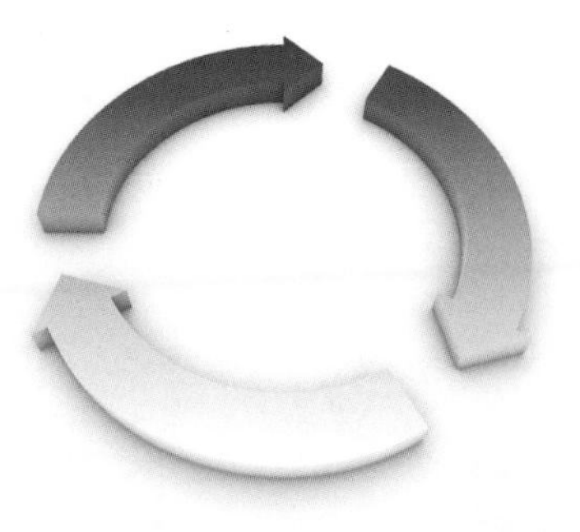

Enlargements

Enlargements change the size but not the shape of the object, i.e. the enlarged shape is **similar** to the object.

The **centre of enlargement** is the point from which the enlargement takes place.

The **scale factor** tells you what all lengths of the original figure have been multiplied by.

An enlargement with a scale factor between 0 and 1 makes the shape smaller.

Example

Describe fully the transformation that maps ABC onto $A'B'C'$.

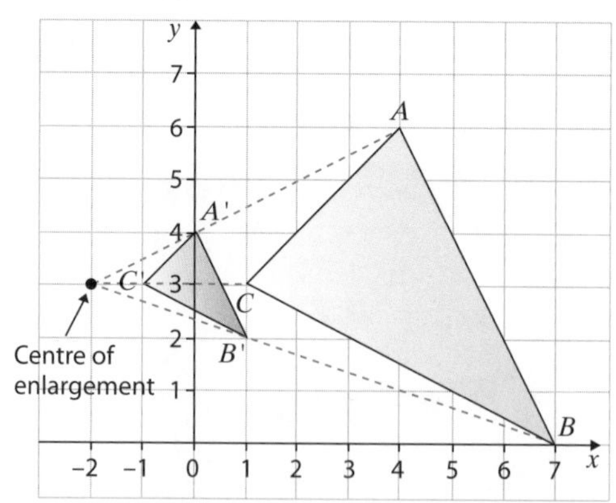

- The transformation is an enlargement.
- To find the centre of enlargement, join A to A', B to B', etc. and continue the lines.
- The centre of enlargement is where all the lines meet: (–2, 3).
- Length $A'B'$ is one-third of the length of AB.

The transformation is an enlargement by scale factor $\frac{1}{3}$. Centre of enlargement is (–2, 3).

Congruent Shapes

Shapes are **congruent** if they are exactly the same size and shape, i.e. they are identical. Two shapes are congruent even if they are mirror images of each other.

Triangles are congruent if one of the following sets of conditions is true:

Condition	Description	Example
SSS (Side–Side–Side)	The three sides of one triangle are the same as the three sides of the other triangle.	
SAS (Side–Angle–Side)	Two sides and the angle between them in one triangle are equal to two sides and the angle between them in the other triangle.	
RHS (Right Angle–Hypotenuse–Side)	Each triangle contains a right angle. Both hypotenuses and another pair of sides are equal.	
AAS (Angle–Angle–Side)	Two angles and a side in one triangle are equal to two angles and the corresponding side in the other.	

SUMMARY

- **After a rotation, the image is congruent to the object.**
- **After an enlargement, the enlarged shape is similar to the object.**
- **An enlargement with a scale factor between 0 and 1 makes the shape smaller.**
- **Shapes are congruent if they are exactly the same size and shape (i.e. they are identical).**

QUESTIONS

QUICK TEST

1. Complete the diagram below, to show the enlargement of the shape by a scale factor of $\frac{1}{2}$. Centre of enlargement at (0, 0). Call the shape T.

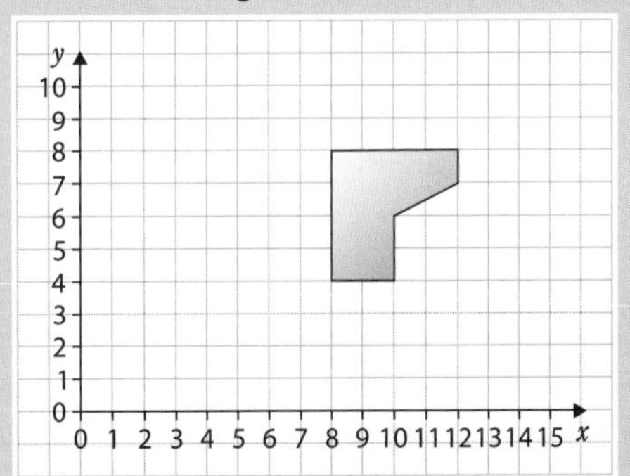

2. Rotate triangle ABC 90° anticlockwise about the point (0, 1). Call the triangle $A'B'C'$.

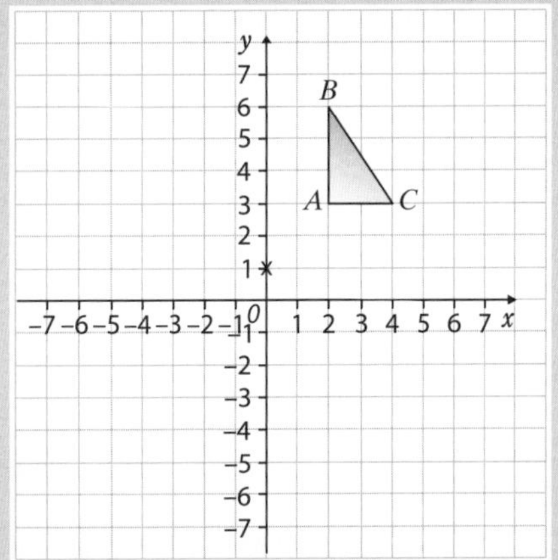

EXAM PRACTICE

1. CDE is an equilateral triangle. F lies on DE. CF is perpendicular to DE.

   Prove that triangle CFD is congruent to triangle CFE. [3 marks]

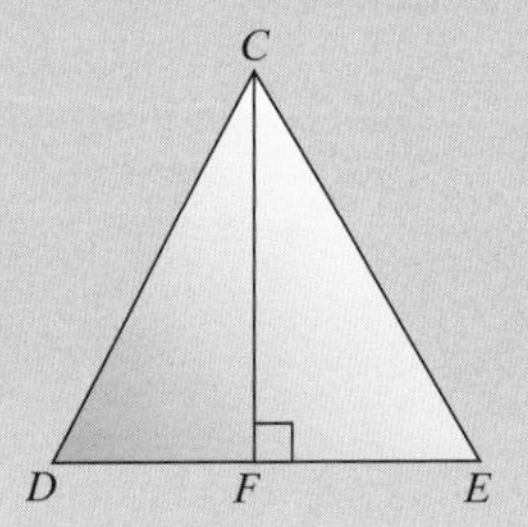

Pythagoras' Theorem

Pythagoras' Theorem states that 'For any right-angled triangle the square on the hypotenuse is equal to the sum of the squares on the other two sides.'

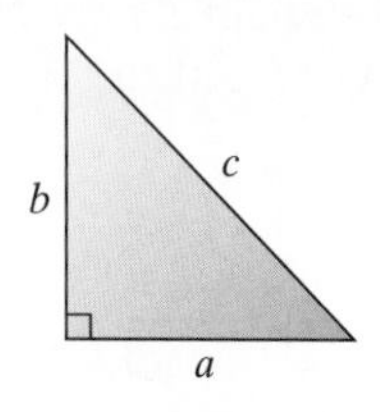

$$a^2 + b^2 = c^2$$

Finding the Hypotenuse

Example

Find the hypotenuse in the right-angled triangle.

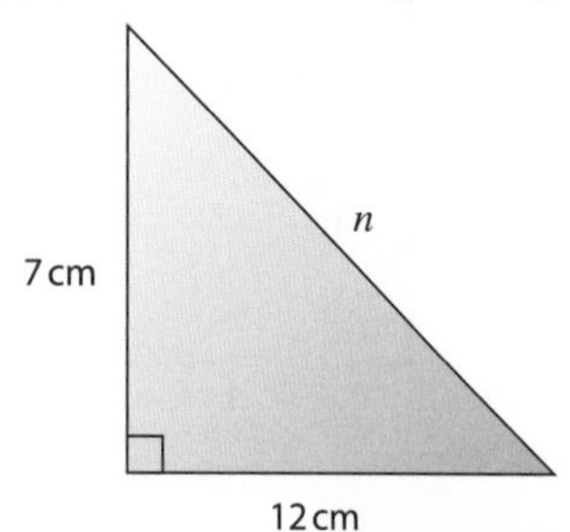

$n^2 = 7^2 + 12^2$ ← Square the two sides.

$n^2 = 49 + 144$ ← Add the two sides together.

$n^2 = 193$

$n = \sqrt{193}$ ← Square root.

$n =$ **13.9 cm** (3 s.f.) ← Round to 3 s.f.

On the non-calculator paper you would give this answer as $\sqrt{193}$ cm.

Finding a Shorter Side

Example

Find the length of p.

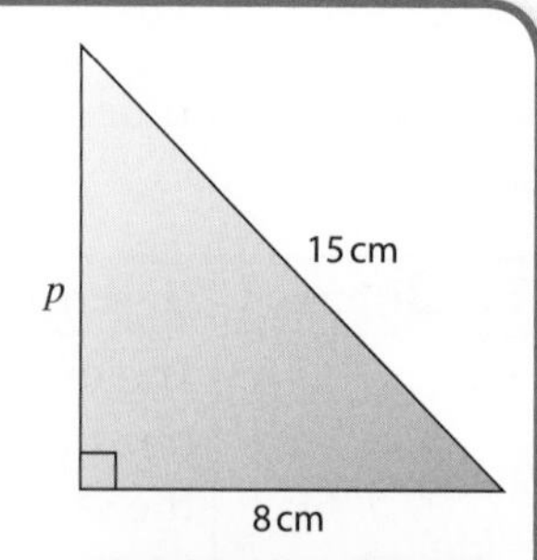

$15^2 = p^2 + 8^2$

$15^2 - 8^2 = p^2$

$225 - 64 = p^2$ ← When finding a shorter length, remember to subtract.

$161 = p^2$

$\sqrt{161} = p$

$p =$ **12.7 cm** (3 s.f.)

Finding the Length of a Line Segment AB, Given the Coordinates of its End Points

Example

Find the length AB in the following diagram.

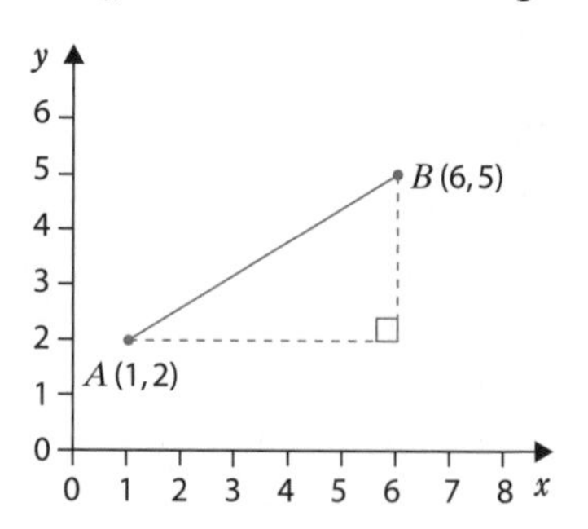

Horizontal distance = 6 – 1 = 5

Vertical distance = 5 – 2 = 3

$AB^2 = 5^2 + 3^2$

$= 25 + 9$

$= 34$

$AB = \sqrt{34}$

$AB =$ **5.83 units** (3 s.f.)

You could leave this as $\sqrt{34}$ units.

Solving a More Difficult Problem

Example

Calculate the vertical height of this isosceles triangle.

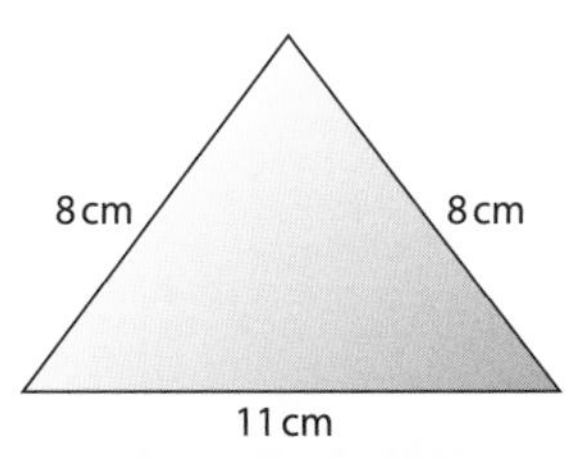

Remember to split the triangle down the middle to make it right-angled.

Using Pythagoras' Theorem gives:

8 cm, h, 5.5 cm

$$8^2 = h^2 + 5.5^2$$

$$64 = h^2 + 30.25$$

$$64 - 30.25 = h^2$$

$$33.75 = h^2$$

$$\sqrt{33.75} = h$$

$h =$ **5.81 cm** (3 s.f.)

SUMMARY

- **Pythagoras' Theorem states that 'For any right-angled triangle, the square on the hypotenuse is equal to the sum of the squares on the other two sides'.**
- **Remember:**

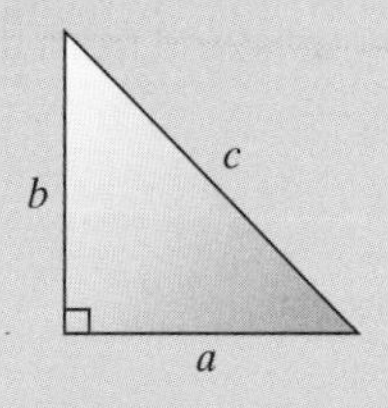

$$a^2 + b^2 = c^2$$

QUESTIONS

QUICK TEST

1. Work out the missing lengths labelled x in the diagrams below. Give your answers to 2 decimal places.

a.

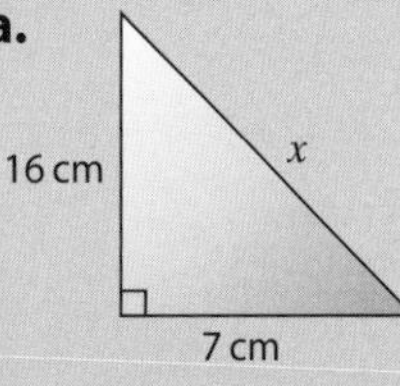

b.

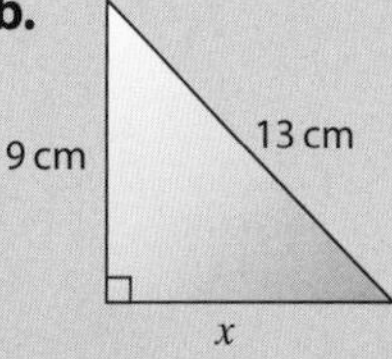

EXAM PRACTICE

1. Molly says: 'The angle x in this triangle is 90°'.

   Explain how Molly knows that without measuring the size of the angle. [2 marks]

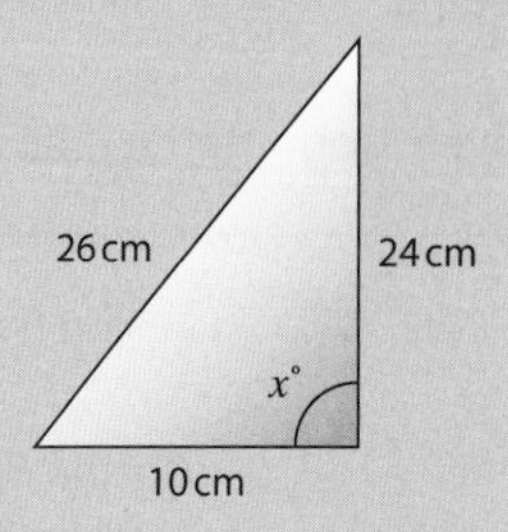

2. The diagram shows a room. Laminate flooring has been laid in the room. Laminate beading is now being placed along the walls of the room. Beading comes in 2.5 metre lengths and costs £1.74 per length.

   Calculate the cost of the beading for the room. [4 marks]

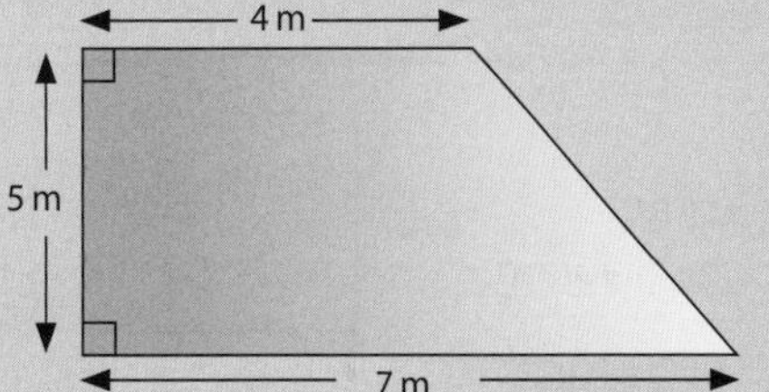

3. Calculate the length of CD in this diagram. Give your answer to 1 decimal place. [2 marks]

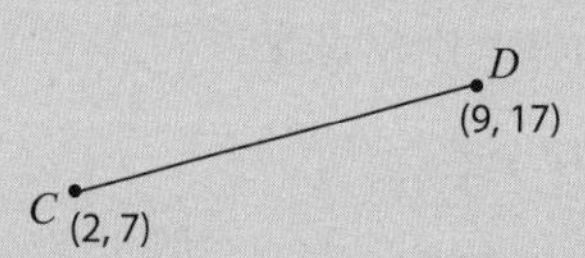

Trigonometry

Trigonometry in right-angled triangles can be used to find an unknown angle or length.

The sides of a right-angled triangle are given temporary names according to where they are in relation to a chosen angle θ.

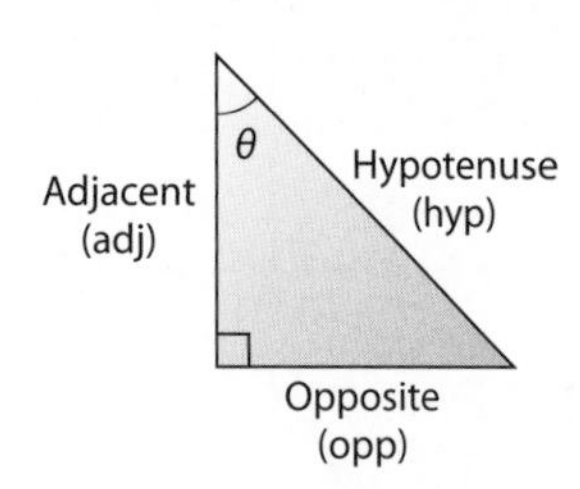

The trigonometric ratios are:

$$\text{Sine }\theta = \frac{\text{Opposite}}{\text{Hypotenuse}}$$

$$\text{Cosine }\theta = \frac{\text{Adjacent}}{\text{Hypotenuse}}$$

$$\text{Tangent }\theta = \frac{\text{Opposite}}{\text{Adjacent}}$$

Use **SOH – CAH – TOA** to remember the ratios.

Example: TOA means $\tan\theta = \frac{\text{opp}}{\text{adj}}$

Finding a Length

Example

Find the missing length y in the diagram.

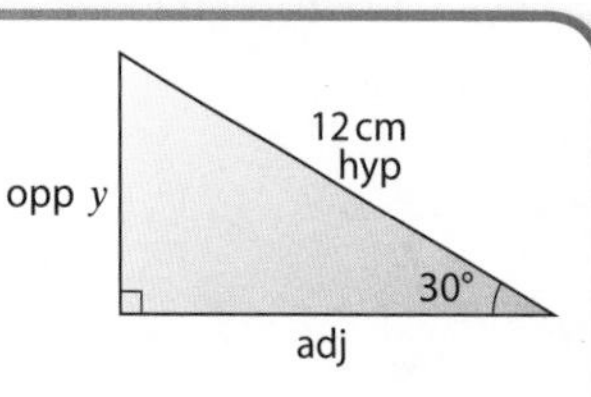

- Label the sides first.
- Decide on the ratio.

$\sin 30° = \frac{\text{opp}}{\text{hyp}}$

- Substitute in the values.

$\sin 30° = \frac{y}{12}$

$12 \times \sin 30° = y$ ← Multiply both sides by 12.

$y =$ **6 cm**

Finding an Angle

Example

Calculate angle ABC.

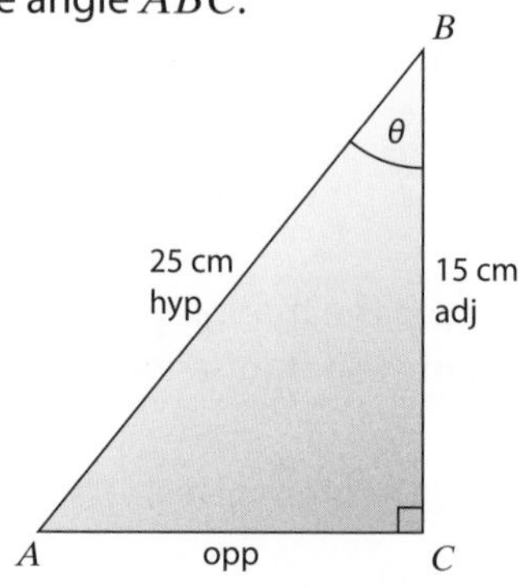

Label the sides and decide on the ratio.

$\cos\theta = \frac{\text{adj}}{\text{hyp}}$

$\cos\theta = \frac{15}{25}$

$\theta = \cos^{-1}\left(\frac{15}{25}\right)$ ← To find the angle, you usually use the second function on your calculator.

$=$ **53.13°** (2 d.p.)

It is important you know how to use your calculator when working out trigonometry questions. Check your calculator is set to work in degrees (not radians).

Angle of Elevation

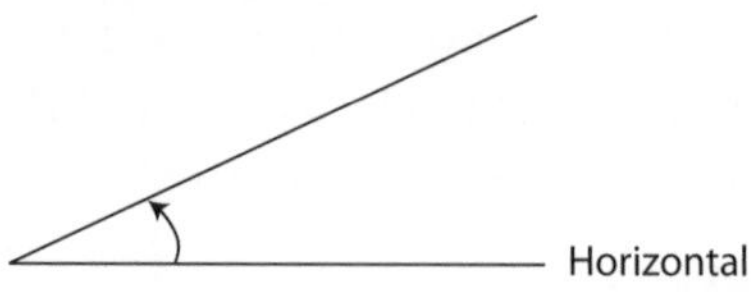

The angle of elevation is the angle measured from the horizontal upwards.

Angle of Depression

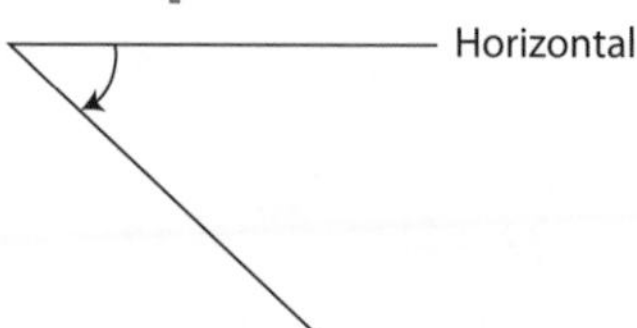

The angle of depression is the angle measured from the horizontal downwards.

Exact Values

The sine, cosine and tangent of some angles can be written exactly.

	0°	30°	45°	60°	90°
sin θ	0	$\frac{1}{2}$	$\frac{\sqrt{2}}{2}$	$\frac{\sqrt{3}}{2}$	1
cos θ	1	$\frac{\sqrt{3}}{2}$	$\frac{\sqrt{2}}{2}$	$\frac{1}{2}$	0
tan θ	0	$\frac{\sqrt{3}}{3}$	1	$\sqrt{3}$	

Example
Find the exact value of x in this triangle.

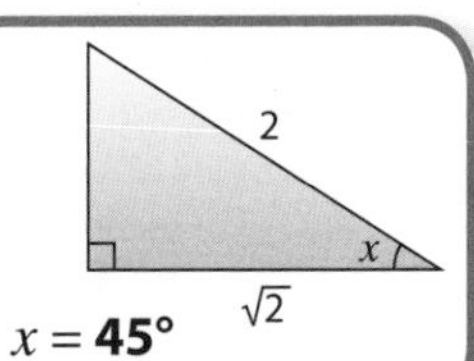

$\cos x = \frac{\text{adj}}{\text{hyp}}$ $\cos x = \frac{\sqrt{2}}{2}$ $x = \mathbf{45°}$

SUMMARY

- **Remember SOH – CAH – TOA:**

$$\textbf{Sine } \theta = \frac{\textbf{Opposite}}{\textbf{Hypotenuse}}$$

$$\textbf{Cosine } \theta = \frac{\textbf{Adjacent}}{\textbf{Hypotenuse}}$$

$$\textbf{Tangent } \theta = \frac{\textbf{Opposite}}{\textbf{Adjacent}}$$

- **The angle of elevation is measured from the horizontal upwards.**
- **The angle of depression is measured from the horizontal downwards.**

QUESTIONS

QUICK TEST

1. Work out the missing lengths labelled x in the diagrams below. Give your answers to 2 decimal places.

a.

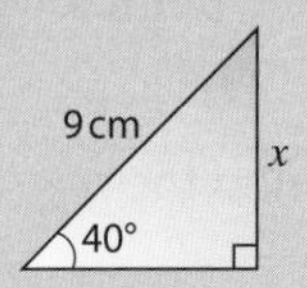

b.

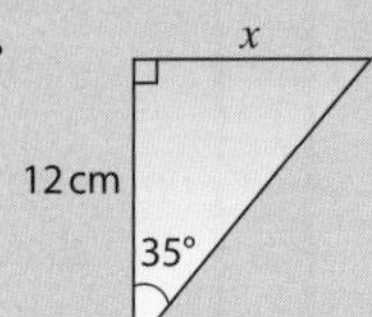

2. Work out the missing angles labelled x in the diagrams below. Give your answers to 1 decimal place.

a.

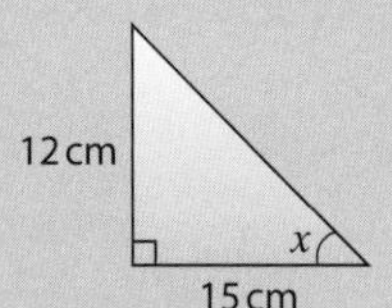

b.

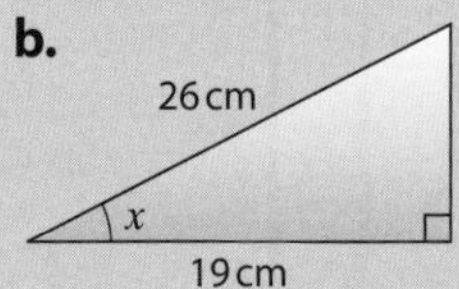

EXAM PRACTICE

1. The diagram represents a vertical mast, PN. The mast is supported by two metal cables, PA and PB, fixed to the horizontal ground at A and B.

$BN = 12.6$ m

$PN = 19.7$ m

angle $PAN = 48°$

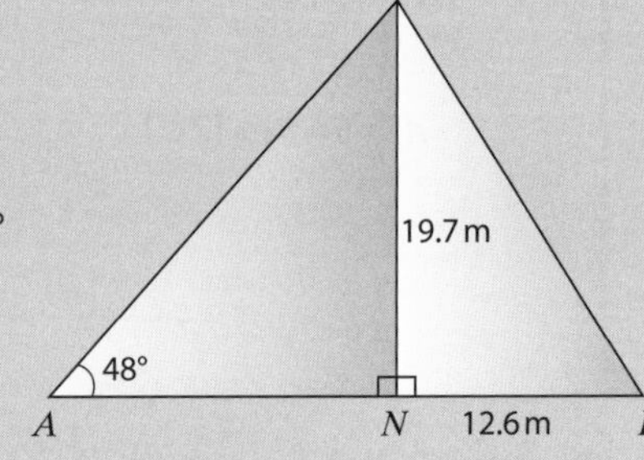

a. Calculate the size of angle PBN. Give your answer correct to 3 significant figures. [3 marks]

b. Calculate the length of the metal cable PA. Give your answer correct to 3 significant figures. [3 marks]

Trigonometry and Pythagoras Problems

On the GCSE paper there will usually be a question which involves you applying **trigonometry** or **Pythagoras' Theorem**, or both. Here are some worked examples.

Example 1

The diagram shows a triangle ABC.

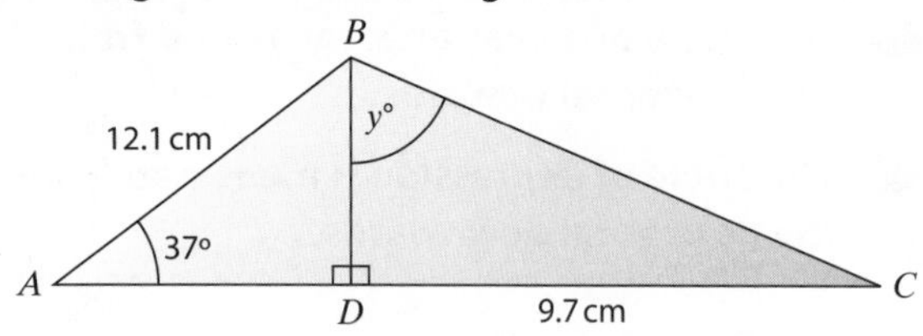

Calculate the size of the angle marked $y°$ if $AB = 12.1$ cm, $CD = 9.7$ cm, $B\hat{A}D = 37°$. Give your answer correct to 1 decimal place.

This is an example of a multi-stepped question. In other words, you have to do several parts before you get to the final answer.

- Find the length of BD first.

$$\sin 37° = \frac{\text{opp}}{\text{hyp}} = \frac{BD}{12.1}$$

$$BD = \sin 37° \times 12.1$$

$$BD = 7.28196\ldots \text{ cm}$$

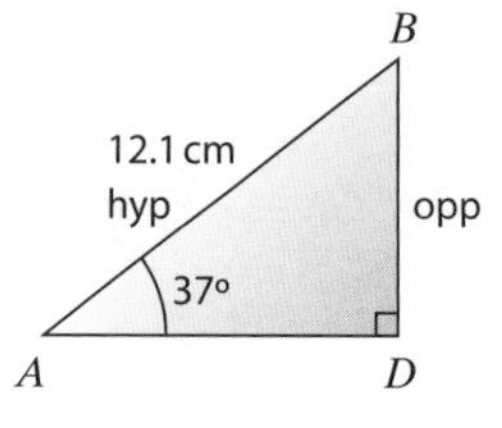

- Now find the angle $y°$.

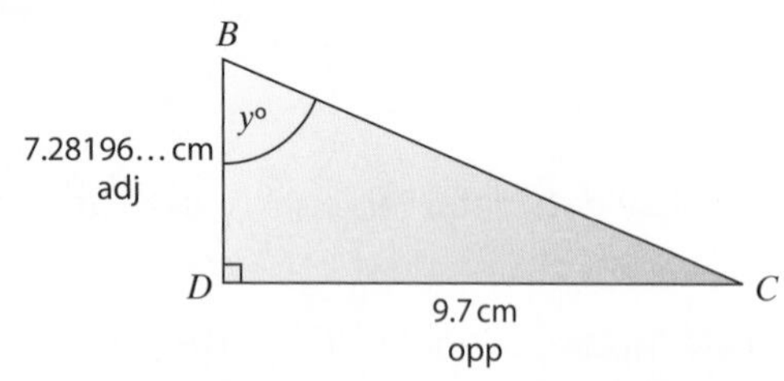

$$\tan y° = \frac{\text{opp}}{\text{adj}} = \frac{9.7}{7.28196\ldots}$$

$$y° = \tan^{-1}(1.33205\ldots)$$

$$y° = \mathbf{53.1°} \text{ (1 d.p.)}$$

Remember not to round this number off until the end.

Example 2

A ship sails 37 km due north and then 42 km due east.

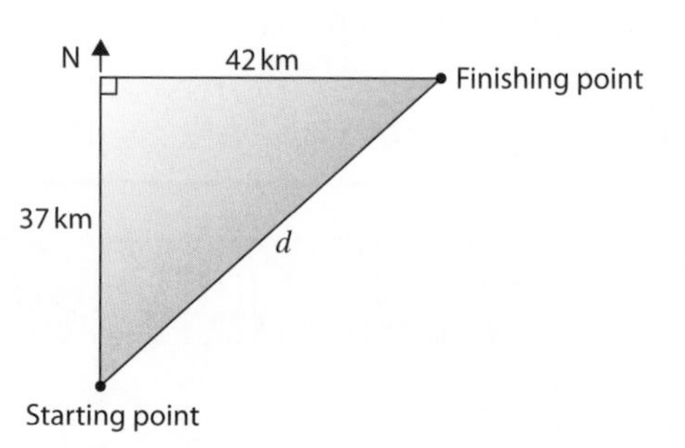

a. Calculate the direct distance between the starting point and the finishing point. Give your answer to 2 significant figures.

Use Pythagoras' Theorem to find the distance (d):

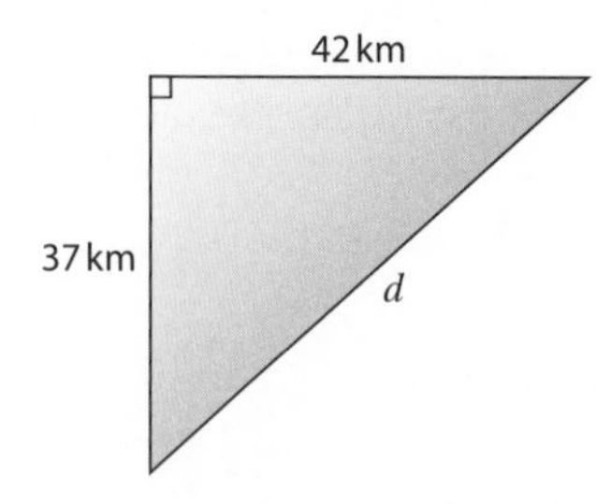

$$d^2 = 37^2 + 42^2$$

$$d^2 = 1369 + 1764$$

$$d^2 = 3133$$

$$d = \sqrt{3133}$$

$$d = \mathbf{56\ km} \text{ (2 s.f.)}$$

Example 2 (cont.)

b. Calculate the bearing of the ship from its starting point. Give your answer to the nearest degree.

Use trigonometry to find the bearing:

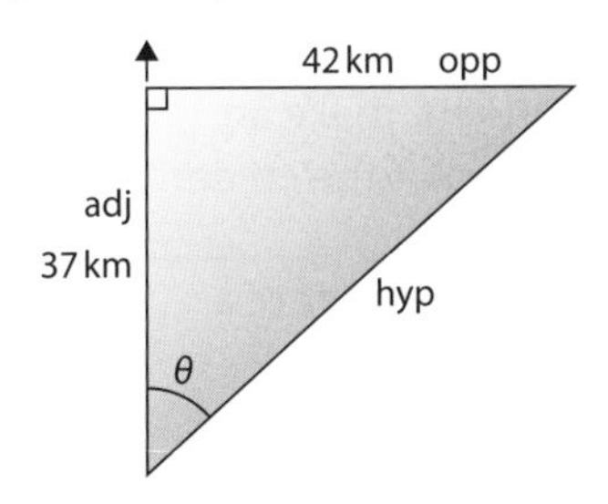

$\tan\theta = \frac{\text{opp}}{\text{adj}} = \frac{42}{37}$

$\theta = \tan^{-1}(1.135\ldots)$ ← Do not round this value.

$\theta = 48.621\ldots°$

Bearing = **049°** (to the nearest degree)

Remember, bearings must be three figures.

SUMMARY

- **Pythagoras' Theorem:** $a^2 + b^2 = c^2$
- **The three trigonometric ratios are:**

$$\textbf{Sine } \theta = \frac{\textbf{Opposite}}{\textbf{Hypotenuse}}$$

$$\textbf{Cosine } \theta = \frac{\textbf{Adjacent}}{\textbf{Hypotenuse}}$$

$$\textbf{Tangent } \theta = \frac{\textbf{Opposite}}{\textbf{Adjacent}}$$

- **Bearings are measured from the North in a clockwise direction; they are written using three figures.**

QUESTIONS

QUICK TEST

1. Work out the area of the triangle. Give your answer to 3 significant figures.

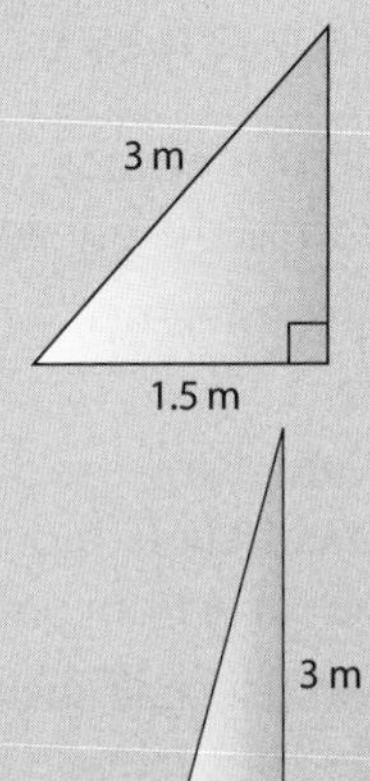

EXAM PRACTICE

1. A ladder is leaning against a wall. Its foot is 0.6 m from the wall and it reaches to a height of 3 m up the wall.

   Calculate the length in metres of the ladder. Give your answer to 2 decimal places. [3 marks]

2. The diagram shows a roof truss.

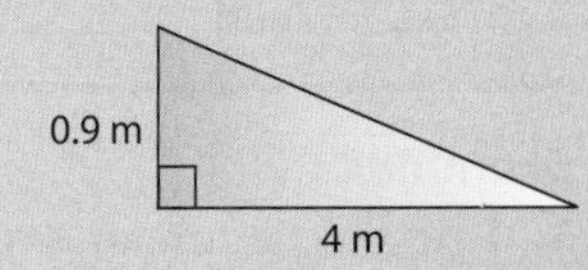

   a. What angle will the roof truss make with the horizontal? Give your answer to 1 decimal place. [3 marks]

   b. What is the length of the sloping strut? Give your answer to 1 decimal place. [3 marks]

3. A, B and C are three towns.

   A is 7.6 km due west of B.
   C is 9.8 km due north of B.

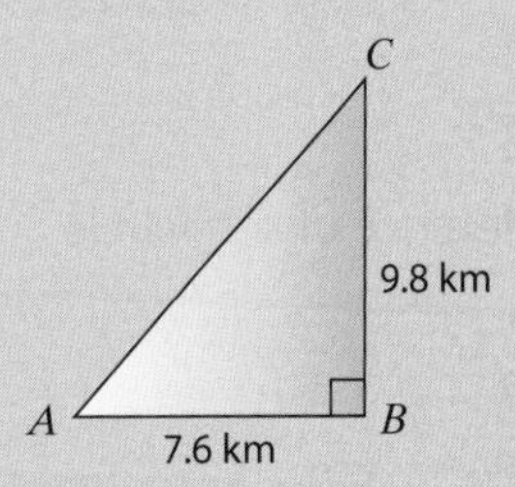

   Calculate the bearing and distance of town A from town C. Give your answers to 3 significant figures. [6 marks]

Circles, Arcs and Sectors

Properties of Circles

Circumference – the distance around the outside edge of a circle. Circumference = $\pi \times$ diameter $= 2 \times \pi \times$ radius **Diameter** – the distance of a straight line through the centre of the circle from one side to the other. **Radius** – half the diameter. Area of circle = $\pi \times (\text{radius})^2$ 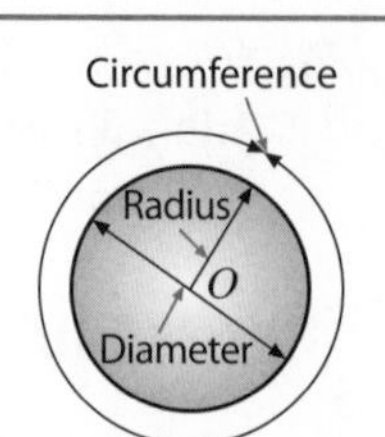	**Tangent** – a line that touches the circle at one point only. The radius and tangent make an angle of 90° at the point they meet. 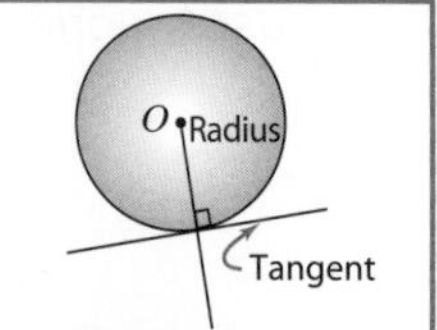
	The **perpendicular bisector** of a chord passes through the centre of a circle. 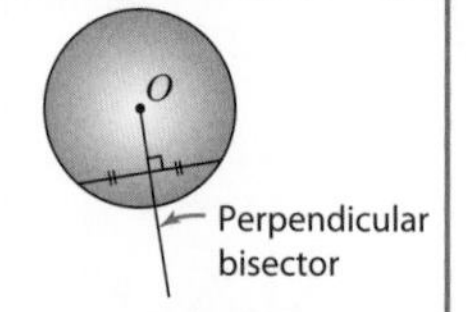
Chord – a line joining two points on the circumference without going through the centre of the circle. **Arc** – part of the circumference.	The angle in a semicircle is always a right angle. 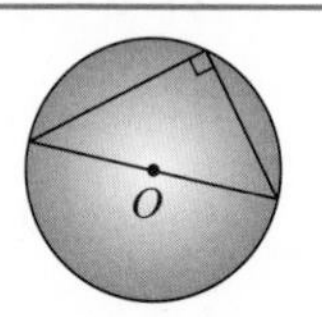

Examples

1. Find the perimeter and area of this shape.

Use the π button on your calculator. Give your answers to 3 significant figures.

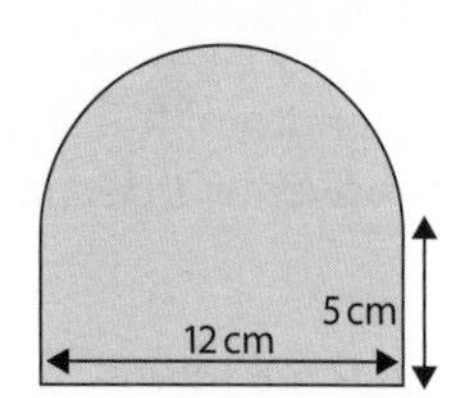

a. Length of straight sides:

$P = 12 + 5 + 5$
$= 22$ cm

Circumference of semicircle:

$C = \dfrac{\pi \times 12}{2}$
$= 18.849\ldots$ cm

Perimeter of shape:

$= P + C$
$= 22 + 18.849\ldots$
$= 40.849\ldots$
$=$ **40.8 cm** (3 s.f.)

b. Area of rectangle $= l \times w$
$= 12 \times 5$
$= 60$ cm^2

Area of semicircle $= \dfrac{\pi \times r^2}{2}$
$= \dfrac{\pi \times 6^2}{2}$
$= 56.548\ldots$ cm^2

Total area $= 60 + 56.548\ldots$
$= 116.548\ldots$
$=$ **117 cm²** (3 s.f.)

2. A circle has an area of 60 cm^2. Find the radius of the circle, giving your answer to 3 significant figures. Use $\pi = 3.142$

$A = \pi \times r^2$

$60 = 3.142 \times r^2$ ← Substitute the values into the formula.

$\dfrac{60}{3.142} = r^2$ ← Divide both sides by 3.142

$r^2 = 19.0961\ldots$

$r = \sqrt{19.0961\ldots}$ ← Take the square root to find r.

$r =$ **4.37 cm** (3 s.f.)

Length of a Circular Arc

The length of a circular **arc** can be expressed as a fraction of the **circumference** of a circle.

$$\text{Arc length} = \frac{\theta}{360^\circ} \times 2\pi r$$

where θ is the angle subtended at the centre

Example

Work out the length of the minor arc AB.

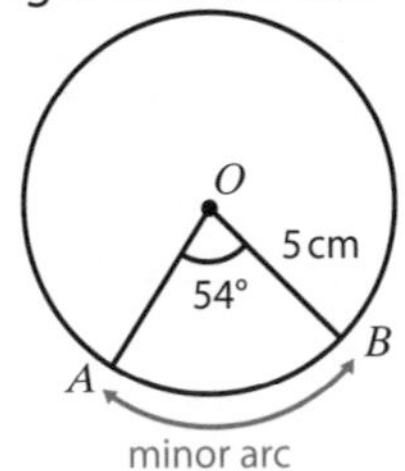

$= \frac{54^\circ}{360^\circ} \times 2 \times \pi \times 5$ $\quad O$ is the centre of the circle.

$= \frac{3}{2}\pi$ **cm** (non-calculator paper) or

4.71 cm (2 d.p.) (calculator paper)

Area of a Sector

The **area** of a **sector** can be expressed as a fraction of the area of a circle.

$$\text{Area of a sector} = \frac{\theta}{360^\circ} \times \pi r^2$$

where θ is the angle subtended at the centre

Example

Work out the area of the minor sector AOB.

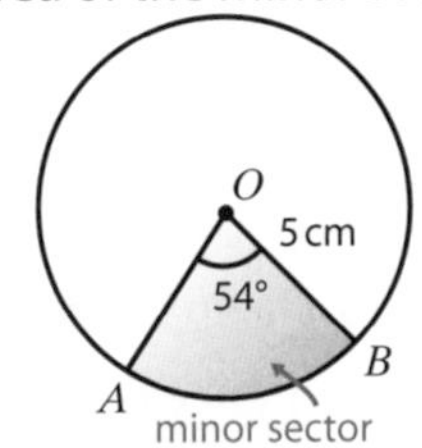

$= \frac{54^\circ}{360^\circ} \times \pi \times 5^2$

$= \frac{15}{4}\pi$ **cm²** (non-calculator paper) or

11.78 cm² (2 d.p.) (calculator paper)

SUMMARY

- **The circumference of a circle is also its perimeter.**
- **Circumference of a circle: $C = \pi d$ or $2\pi r$**

  Area of a circle: $A = \pi r^2$
- **Arc length $= \frac{\theta}{360^\circ} \times 2\pi r$**
- **Area of a sector $= \frac{\theta}{360^\circ} \times \pi r^2$**

QUESTIONS

QUICK TEST

1. Find the circumference and area of these circles. Use $\pi = 3.142$

   Give your answers to 2 decimal places.

   a. 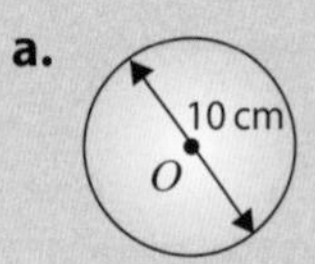

   b.

   c. 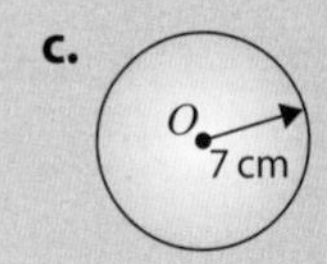

2. Find the perimeter and area of this shape.

   Use the π button.

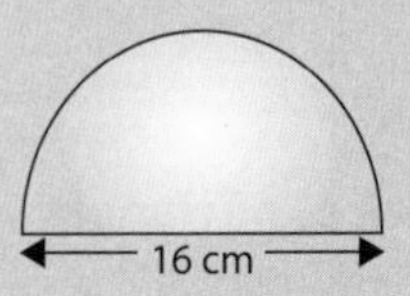

3. For this diagram, calculate:

   a. the arc length (x)

   b. the sector area, AOB

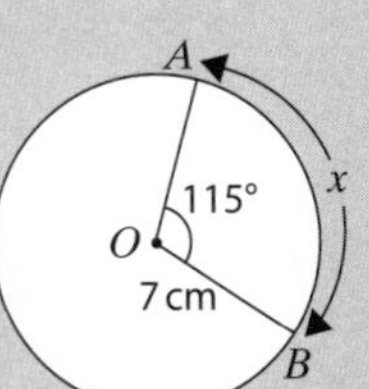

EXAM PRACTICE

1. The diagram shows the inside of a running track.

   The groundsperson is going to cover this area with grass seed. One sack of grass seed covers 275 m².

   How many sacks of grass seed does the groundsperson need? [4 marks]

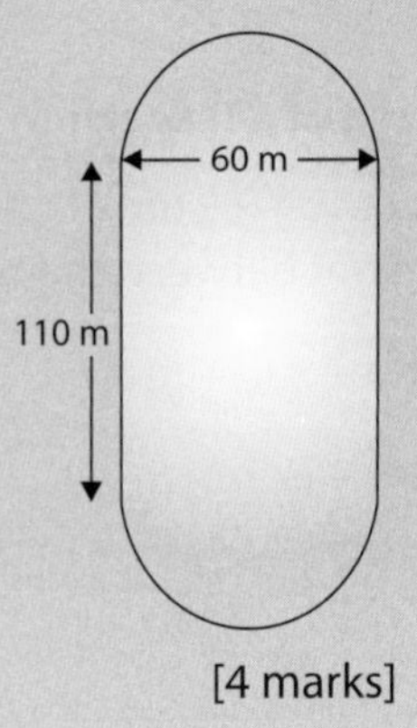

Surface Area and Volume 1

Areas of Plane Shapes

The area of a 2D shape is the amount of flat space that it covers. Common units of area are mm^2, cm^2, m^2.

Key Formulae

Area of a Rectangle

Area = length × width

$A = l \times w$

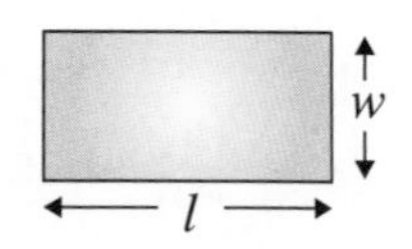

Area of a Parallelogram

Area = base × perpendicular height

$A = b \times h$

Remember to use the perpendicular height, not the slant height.

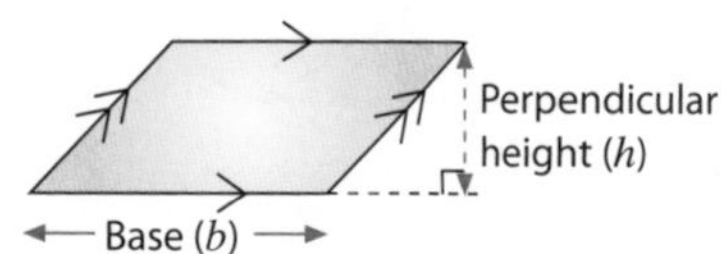

Area of a Triangle

Area = $\frac{1}{2}$ base × perpendicular height

$A = \frac{1}{2} \times b \times h$

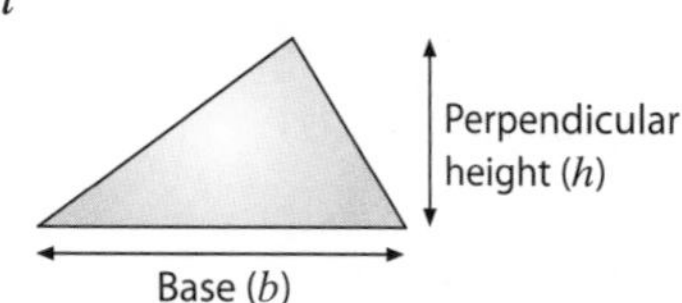

Area of a Trapezium

Area = $\frac{1}{2}$ × (sum of parallel sides) × perpendicular height between them

$A = \frac{1}{2} \times (a + b) \times h$

$A = \frac{1}{2} (a + b) h$

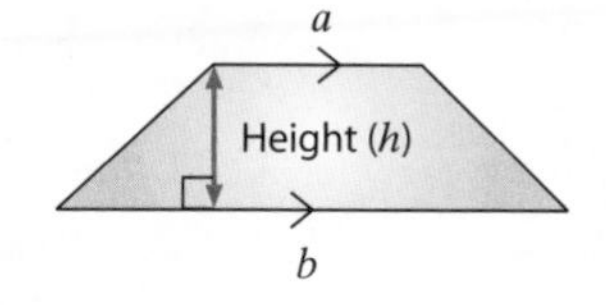

Area of Complex Shapes

Work out the area of more complex shapes by splitting them into simple shapes.

Example

The diagram below shows the plan of a garden.

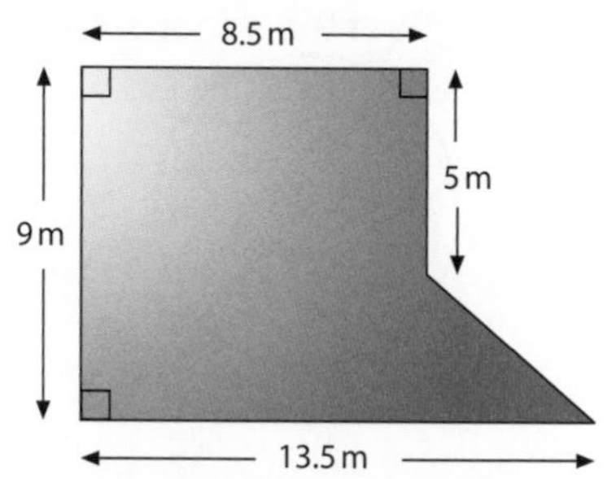

Lawn seed is sown to cover the garden. Lawn seed comes in 500 g packets and covers 15 m^2. A packet of lawn seed costs £6.25. Work out the total cost of the lawn seed needed.

Work out the area of the garden by splitting it into a rectangle and a triangle.

Area of rectangle = $l \times w$

$= 9 \times 8.5$

$= 76.5\ m^2$

Area of triangle = $\frac{1}{2} \times b \times h$

$= \frac{1}{2} \times 5 \times 4$

$= 10\ m^2$

Base of triangle = 13.5 – 8.5 = 5 m
Height of triangle = 9 – 5 = 4 m

Total area = 76.5 + 10

$= 86.5\ m^2$

Number of packets of lawn seed needed:

$\frac{86.5}{15} = 5.76\ldots$ packets, hence 6 packets of lawn seed are needed.

Cost = £6.25 × 6

= **£37.50**

Volumes of Prisms

A **prism** is any solid that can be cut up into slices that are all the same shape. This is known as having a uniform **cross-section**.

Volume of a Cuboid

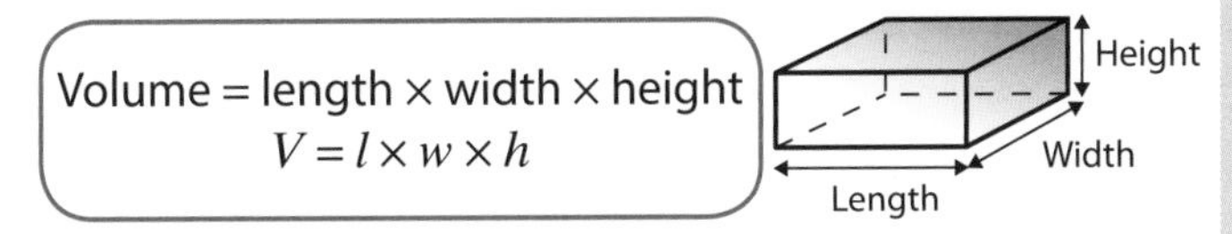

To find the surface area of a cuboid, work out the area of each face, then add together.

Surface area = $2hl + 2hw + 2lw$

Volume of a Prism

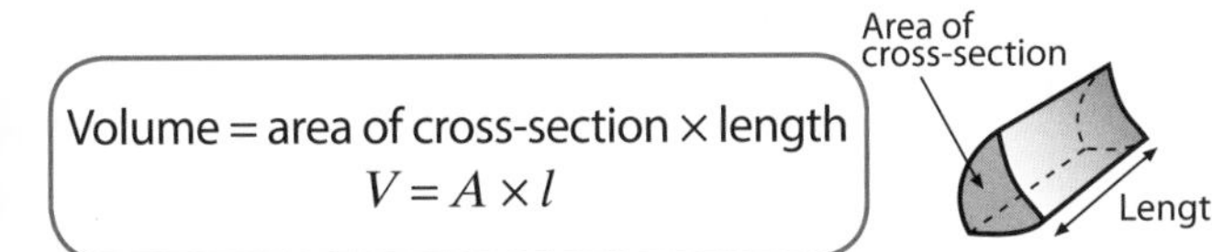

Volume of a Cylinder

Cylinders are prisms in which the cross-section is a circle.

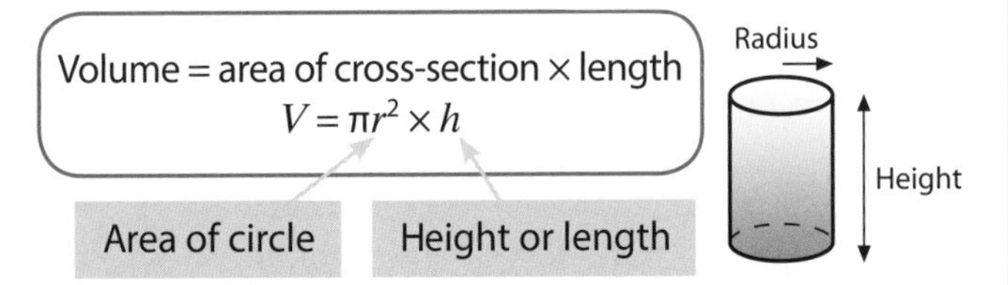

Example

Find the volume of this prism.

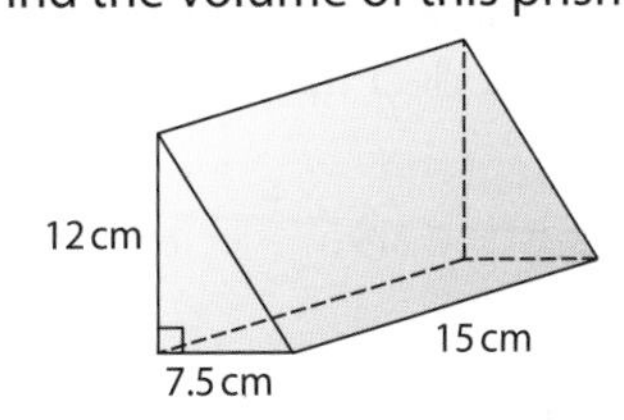

$V = A \times l$

$V = (\frac{1}{2} \times b \times h) \times l$ ← The area of the cross-section is the area of a triangle.

$V = (\frac{1}{2} \times 7.5 \times 12) \times 15$

$V =$ **675 cm³** ← Cubic units for volume

SUMMARY

- **Area is measured in square units.**
- **Area of a rectangle: $A = l \times w$**
  Area of a parallelogram: $A = b \times h$
  Area of a triangle: $A = \frac{1}{2} \times b \times h$
  Area of a trapezium: $A = \frac{1}{2}(a + b)h$
- **Volume is measured in cubic units.**
- **Volume of a cuboid: $V = l \times w \times h$**
  Volume of a prism: $V = A \times l$
  Volume of a cylinder: $V = \pi r^2 \times h$

QUESTIONS

QUICK TEST

1. Find the areas of these shapes:

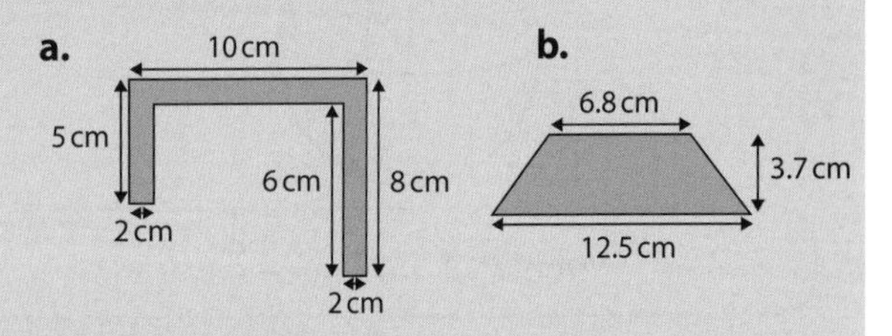

2. Work out the volumes of these prisms. Give your answers to 1 d.p. Use π = 3.142

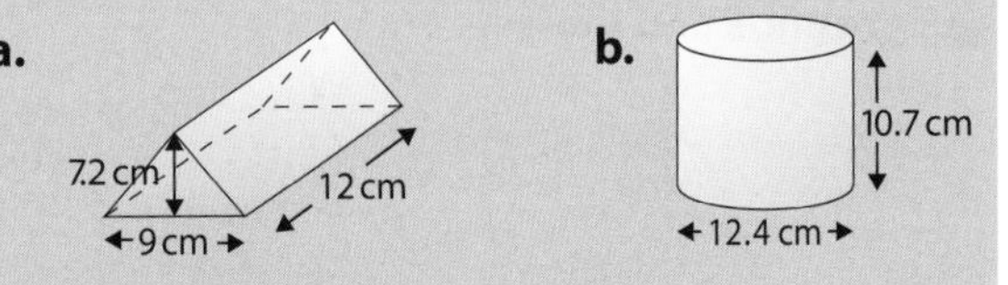

EXAM PRACTICE

1. The diagram shows the plan of a room. Under-floor heating is being installed in the room.
   1 m² of under-floor heating costs £155.

   4 m
   3 m
   6 m

   Work out the total cost of installing under-floor heating for the whole room. [4 marks]

Surface Area and Volume 2

Surface Area of a Cylinder

The total **surface area** of a cylinder can be found by drawing a net.

Cylinder | **Net of a Cylinder**

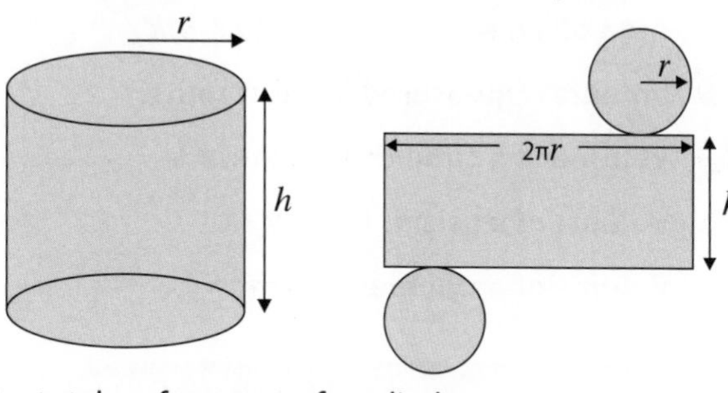

The total surface area of a cylinder =

$2\pi r^2$ (area of the two circles) + $2\pi rh$ (area of the rectangle)

Sphere

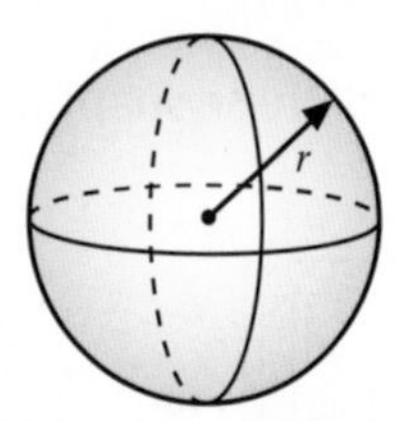

Volume of a sphere $= \frac{4}{3}\pi r^3$

Surface area of a sphere $= 4\pi r^2$

Pyramids and Cones

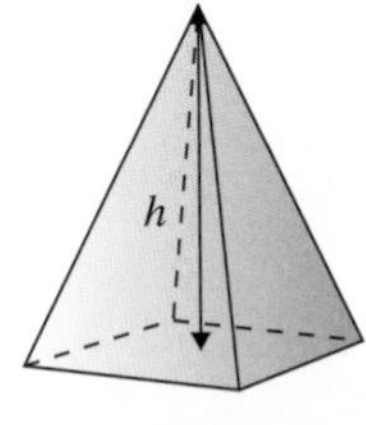

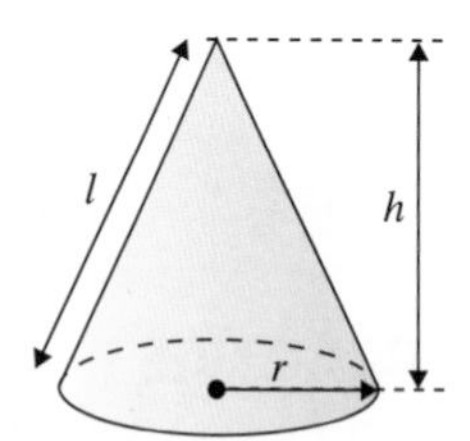

A cone is simply a pyramid with a circular base.

Volume of a pyramid $= \frac{1}{3} \times$ area of base $\times$ height

Volume of a cone $= \frac{1}{3} \times \pi \times r^2 \times$ height

$V = \frac{1}{3}\pi r^2 h$

Curved surface area of cone $= \pi rl$ (l is the slant height)

Examples

1. Find the total surface area of this cylinder. Leave your answer in terms of π.

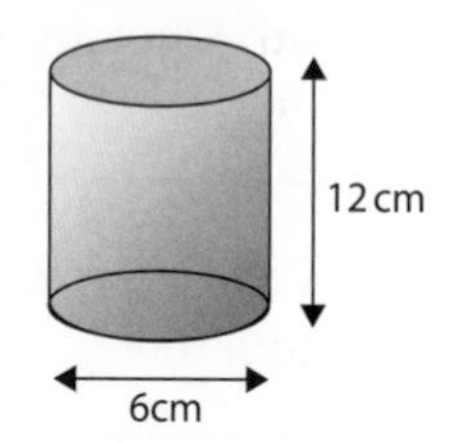

$$SA = 2\pi r^2 + 2\pi rh$$
$$= 2 \times \pi \times 3^2 + 2 \times \pi \times 3 \times 12$$
$$= 18\pi + 72\pi$$
$$= \mathbf{90\pi\ cm^2}$$

2. The diagram below shows a cone.

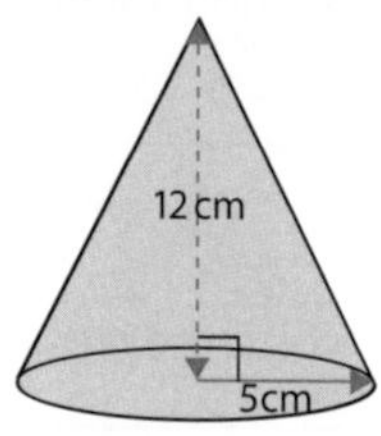

a. Calculate the volume of the cone. Give your answer to 2 decimal places.

$$V = \frac{1}{3}\pi r^2 h$$
$$= \frac{1}{3} \times \pi \times 5^2 \times 12$$
$$= 100\,\pi$$
$$= \mathbf{314.16\ cm^3}\ \text{(2 d.p.)}$$

b. Calculate the total surface area of the cone. Give your answer to 2 decimal places.

Total surface area = curved surface area + area of circle

Slant height, $l = \sqrt{12^2 + 5^2}$
$= \sqrt{144 + 25}$
$= \sqrt{169}$
$= 13$ cm

$$A = \pi rl + \pi r^2$$
$$= \pi \times 5 \times 13 + \pi \times 5^2$$
$$= 65\pi + 25\pi$$
$$= 90\pi$$
$$= \mathbf{282.74\ cm^2}\ \text{(2 d.p.)}$$

Example

The mass of this solid toy is 637 grams.

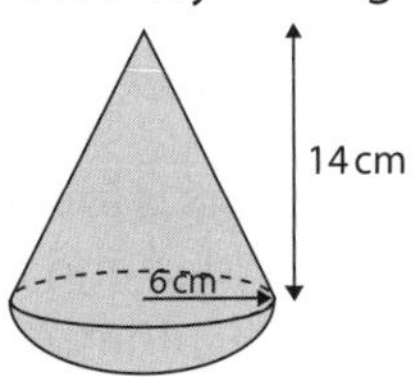

Work out the density of the toy. Give your answer to 2 decimal places.

First, work out the volume of the toy.

Volume of cone $= \frac{1}{3}\pi r^2 h$

$= \frac{1}{3} \times \pi \times 6^2 \times 14$

$= 527.787\ldots \text{ cm}^3$

Volume of sphere $= \frac{4}{3}\pi r^3$

Volume of sphere $= \frac{4}{3} \times \pi \times 6^3$

Volume of hemisphere $= \frac{904.7786\ldots}{2}$

$= 452.389\ldots \text{ cm}^3$

Total volume $= 527.787\ldots + 452.389\ldots$

$= 980.176\ldots \text{ cm}^3$

Density $= \frac{\text{mass}}{\text{volume}}$

Density $= \frac{637}{980.176\ldots}$

Density = **0.65 g/cm³** (2 d.p.)

SUMMARY

- **Volume of a prism = area of cross-section × length**
- **Volume of a sphere = $\frac{4}{3}\pi r^3$**
- **Surface area of a sphere = $4\pi r^2$**
- **Volume of a pyramid = $\frac{1}{3}$ × area of base × height**
- **Volume of a cone = $\frac{1}{3}\pi r^2 h$**
- **Curved surface area of a cone = $\pi r l$ (l is the slant height)**
- **Surface area of a cylinder = $2\pi r^2 + 2\pi r h$**

QUESTIONS

QUICK TEST

1. The volumes of the solids below have been calculated. Match the correct solid with the correct volume.

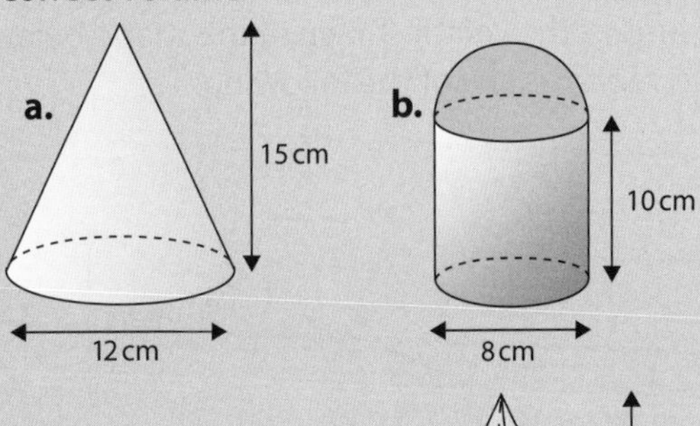

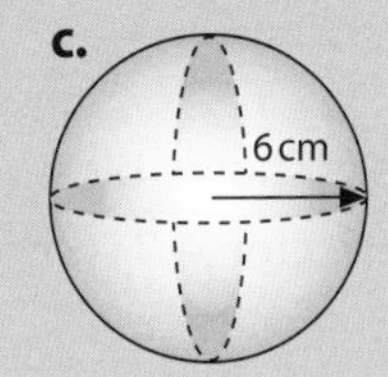

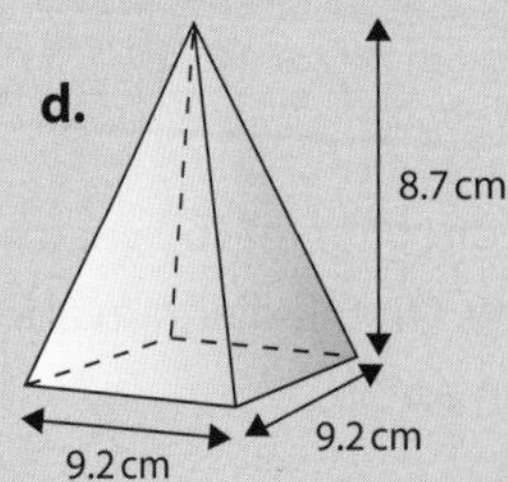

905 cm³ 245 cm³ 565 cm³ 637 cm³

EXAM PRACTICE

1. If the volume of this cylinder is 205 cm³, work out the height. Use π = 3.142 or the π button on your calculator.

   Give your answer to 3 significant figures. [3 marks]

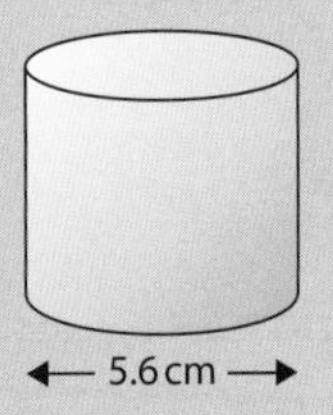

2. A cone has a volume of 15 m³. The vertical height of the cone is 2.1 m.

   Calculate the radius of the base of the cone. Give your answer to 3 significant figures. [4 marks]

Vectors

Vectors

A **vector** is a quantity that has both distance and direction. A **scalar** is a quantity that has only distance. Four types of notation are used to represent vectors. For example, the vector shown in the shape below can be referred to as any of the following:

$\begin{pmatrix} 5 \\ 2 \end{pmatrix}$ $\qquad$ $\underline{a}$ $\qquad$ $\overrightarrow{AB}$ $\qquad$ $\mathbf{a}$

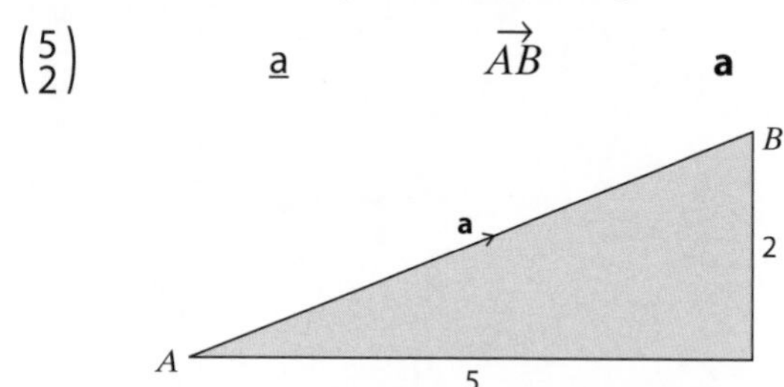

The direction of the vector is usually shown by an arrow. On the diagram above, the vector $\overrightarrow{AB}$ is shown by an arrow.

- If $\overrightarrow{DE} = k\overrightarrow{AB}$, then $\overrightarrow{AB}$ and $\overrightarrow{DE}$ are parallel and the length of $\overrightarrow{DE}$ is k times the length of $\overrightarrow{AB}$.

  $\overrightarrow{AB} = \begin{pmatrix} 5 \\ 2 \end{pmatrix}$

  $\overrightarrow{DE} = \begin{pmatrix} 10 \\ 4 \end{pmatrix} = 2\begin{pmatrix} 5 \\ 2 \end{pmatrix}$

  $\overrightarrow{DE} = 2\overrightarrow{AB}$

- If two vectors are equal, they are parallel and equal in length.

  $\overrightarrow{AB}$ is equal to **a**.

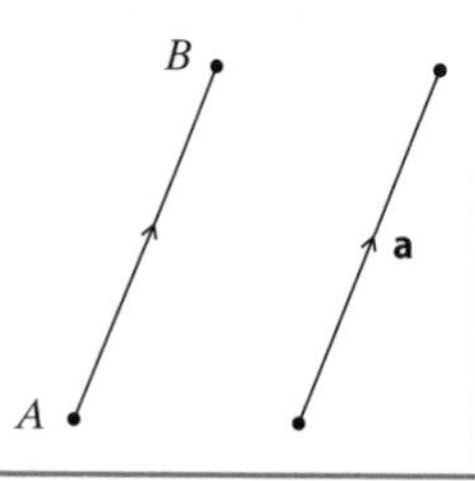

- If the vector $\mathbf{c} = \begin{pmatrix} 3 \\ 4 \end{pmatrix}$

  then the vector **–c** is in the opposite direction to **c**.

  $-\mathbf{c} = \begin{pmatrix} -3 \\ -4 \end{pmatrix}$

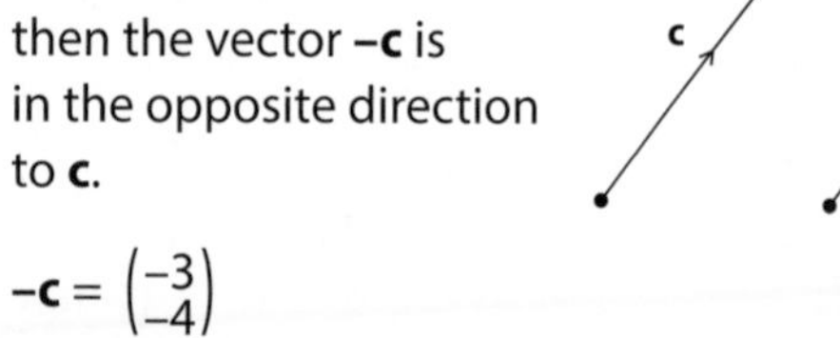

Addition and Subtraction of Vectors

The **resultant** of two vectors is found by adding them. Vectors must always be added end to end so that the arrows follow on from each other. A resultant is usually labelled with a double arrow.

Addition

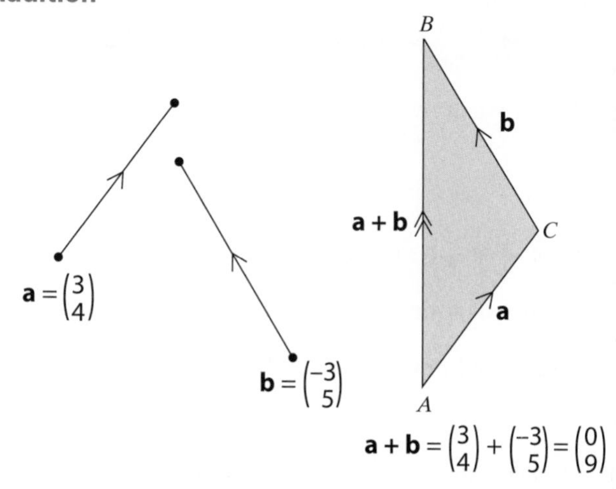

$\mathbf{a} = \begin{pmatrix} 3 \\ 4 \end{pmatrix}$ $\qquad$ $\mathbf{b} = \begin{pmatrix} -3 \\ 5 \end{pmatrix}$

$\mathbf{a} + \mathbf{b} = \begin{pmatrix} 3 \\ 4 \end{pmatrix} + \begin{pmatrix} -3 \\ 5 \end{pmatrix} = \begin{pmatrix} 0 \\ 9 \end{pmatrix}$

To take the route directly from A to B is equivalent to travelling via C. Hence you can represent $\overrightarrow{AB}$ as **a** + **b**.

Subtraction

a – b can be interpreted as **a** + (**–b**).

$\mathbf{a} + (-\mathbf{b}) = \begin{pmatrix} 3 \\ 4 \end{pmatrix} + \begin{pmatrix} 3 \\ -5 \end{pmatrix}$

$\therefore\ \mathbf{a} - \mathbf{b} = \begin{pmatrix} 6 \\ -1 \end{pmatrix}$

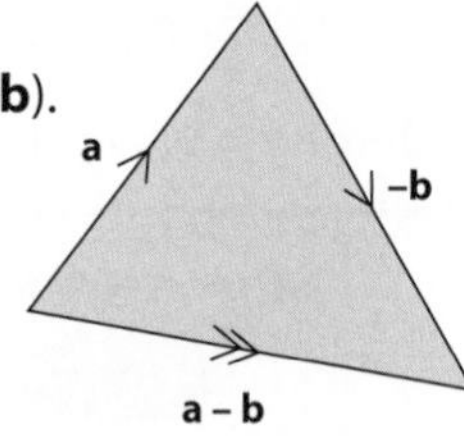

Position Vectors

The position vector of a point B is the vector $\overrightarrow{OB}$, where O is the origin.

In the diagram, the position vectors of B and C are **b** and **c** respectively. Using this notation:

$\overrightarrow{BC} = -\mathbf{b} + \mathbf{c}$

$\quad = \mathbf{c} - \mathbf{b}$

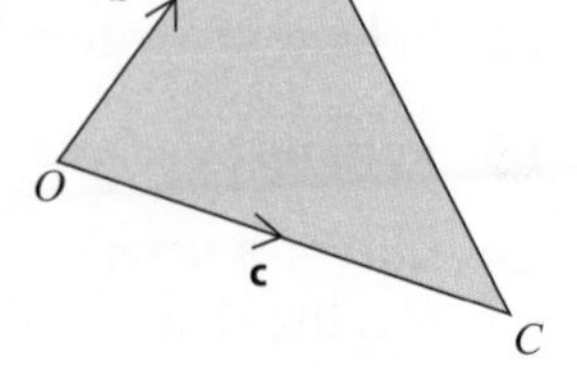

Example

OAB is a triangle. Given that $\overrightarrow{OA} = \mathbf{a}$, $\overrightarrow{OB} = \mathbf{b}$ and that N splits $\overrightarrow{AB}$ in the ratio 1 : 2, find in terms of **a** and **b** the vectors:

i. $\overrightarrow{AB}$ **ii.** $\overrightarrow{ON}$

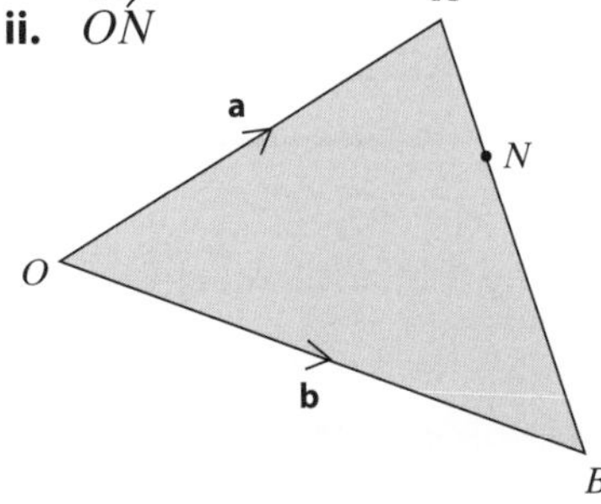

i. $\overrightarrow{AB} = \overrightarrow{AO} + \overrightarrow{OB}$

$= -\mathbf{a} + \mathbf{b}$

Go from A to B via O.

ii. $\overrightarrow{ON} = \overrightarrow{OA} + \overrightarrow{AN}$

$\overrightarrow{AN} = \frac{1}{3}\overrightarrow{AB}$

$= \mathbf{a} + \frac{1}{3}(-\mathbf{a} + \mathbf{b})$

$= \mathbf{a} - \frac{1}{3}\mathbf{a} + \frac{1}{3}\mathbf{b}$

$= \frac{2}{3}\mathbf{a} + \frac{1}{3}\mathbf{b}$

$= \frac{1}{3}(2\mathbf{a} + \mathbf{b})$

SUMMARY

- **The direction of a vector is shown with an arrow.**
- **The resultant of two vectors is found by vector addition or subtraction.**
- **Vectors must always be combined end to end.**

QUESTIONS

QUICK TEST

1. The diagram is a sketch.
A is the point (3, 2)
B is the point (7, 7)

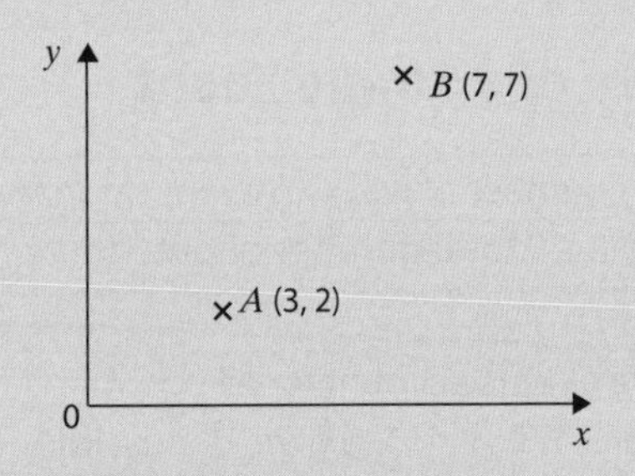

a. Write down the vector $\overrightarrow{AB}$.
Write your answer as a column vector $\begin{pmatrix} x \\ y \end{pmatrix}$.

b. Write down the coordinates of point C such that $\overrightarrow{BC} = \begin{pmatrix} -2 \\ -3 \end{pmatrix}$.

EXAM PRACTICE

1. $OABC$ is a parallelogram.
AB is parallel to OC.
OA is parallel to CB.

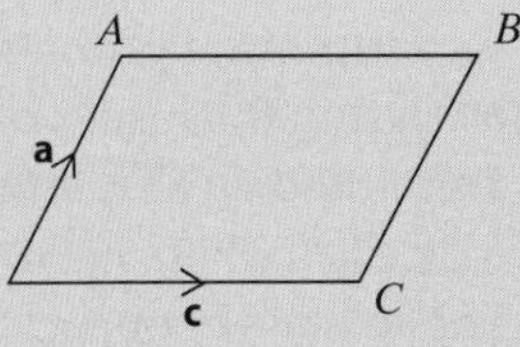

$\overrightarrow{OA} = \mathbf{a}$

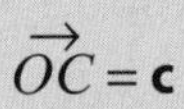

$\overrightarrow{OC} = \mathbf{c}$

a. Express in terms of **a** and **c**:

i. $\overrightarrow{AC}$ [1 mark]

ii. $\overrightarrow{BO}$ [1 mark]

b. N is the midpoint of $\overrightarrow{AC}$:

Express $\overrightarrow{ON}$ in terms of **a** and **c**. [2 marks]

Probability

Probability is the chance or likelihood that something will happen. All probabilities lie between 0 (impossible) and 1 (certain).

Probabilities must be written as a fraction, decimal or percentage.

Probability of a Single Event

$$\text{P(event)} = \frac{\text{number of ways an event can happen}}{\text{total number of outcomes}}$$

- **Exhaustive events** account for all possible outcomes. For example, 1, 2, 3, 4, 5, 6 give all possible outcomes when a dice is thrown.
- **Mutually exclusive** events are events that cannot happen at the same time, e.g. a head and a tail on a fair coin cannot appear at the same time.
- Two or more events are **independent** when the outcome of the second event is not affected by the outcome of the first event.
- **Theoretical probability** analyses a situation mathematically.
- **Experimental probability** is determined by analysing the results of a number of trials or events. This is known as **relative frequency**. For example, a dice is thrown 55 times. A four comes up 13 times. The relative frequency is $\frac{13}{55}$

Probability that an Event will NOT Happen

P(event will not happen) = 1 – P(event will happen)

Example

The probability that an alarm clock fails to go off is 0.21

What is the probability that the alarm clock will go off?

1 – 0.21 = **0.79**

Expected Number

Probability helps you predict the outcome of an event.

The expected number of outcomes = number of trials × probability

Example

The probability of passing an exam in microbiology is 0.37

If 100 people take the exam, how many are expected to pass?

100 × 0.37 = **37 people**

Sample Space Diagrams

Sample space diagrams can be helpful when considering the outcomes of two events.

Example

Two spinners are spun and the scores added.

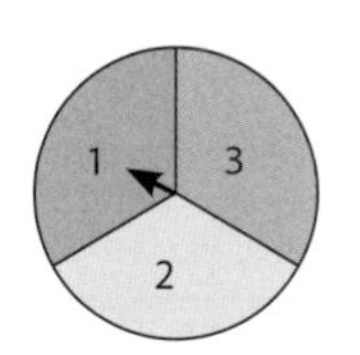

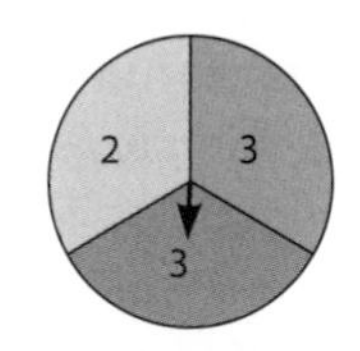

a. Represent the outcomes on a sample space diagram.

		Spinner 1		
		1	2	3
	2	3	4	5
Spinner 2	3	4	5	6
	3	4	5	6

b. What is the probability of scoring 4 in total?

$P(4) = \frac{3}{9} = \mathbf{\frac{1}{3}}$

c. What is the probability of scoring an odd number in total?

$P(\text{odd}) = \mathbf{\frac{4}{9}}$

SUMMARY

- **All probabilities lie between 0 and 1.**
- **P(event) = $\frac{\text{number of ways an event can happen}}{\text{total number of outcomes}}$**
- **P(event will not happen) = 1 – P(event will happen)**
- **Mutually exclusive events cannot happen at the same time.**

QUESTIONS

QUICK TEST

1. The probability that Ahmed does his homework is 0.65

   What is the probability that Ahmed does not do his homework?

2. The probability of passing a business exam is 0.49

   If 600 students sit this exam, how many would you expect to pass?

3. Two fair dice are thrown and their scores are multiplied. By drawing a sample space diagram, what is:

   a. the probability of a score of 6?

   b. the probability of an even score?

EXAM PRACTICE

1. The probability that a factory manufactures a faulty component is 0.03

   Each day 1200 components are manufactured. Estimate the number of faulty components each day. [2 marks]

2. Imran plays a game of throwing a dart at a target. The table shows information about the probability of each possible score.

Score	0	1	2	3	4	5
Probability	0.04	x	$4x$	0.23	0.19	0.28

   Imran is 4 times more likely to score 2 points than to score 1 point.

   Work out the value of x. [3 marks]

Tree Diagrams

Tree diagrams are used to show the possible outcomes of two or more events. There are two rules you need to know first.

- **The OR rule**
  If two events are **mutually exclusive**, the probability of A or B happening is found by adding the probabilities.

P(A or B) = P(A) + P(B)

(These rules also work for more than two events.)

- **The AND rule**
  If two events are **independent**, the probability of A and B happening together is found by multiplying the separate probabilities.

P(A and B) = P(A) × P(B)

Example

A bag contains three red and four blue counters. A counter is taken from the bag at random, its colour is noted and then it is replaced in the bag. A second counter is then taken out of the bag. Draw a tree diagram to illustrate this information.

Remember that the probabilities on the branches leaving each point on the tree add up to 1.

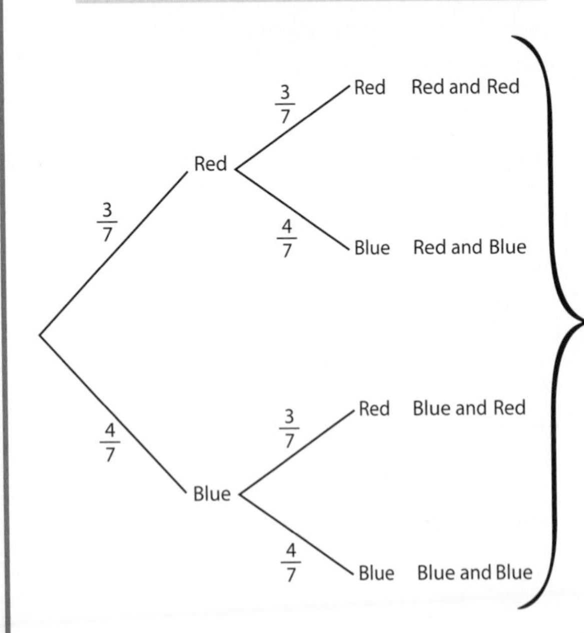

Work out the probability of:

a. picking two blues

To find the probability of picking a blue AND a blue, multiply along the branches.

P(B) × P(B)

$= \frac{4}{7} \times \frac{4}{7} = \mathbf{\frac{16}{49}}$

b. picking one of either colour

The probability of picking one of either colour is the probability of picking a blue and a red OR a red and a blue. You need to use the AND rule and the OR rule.

P(blue and red)

P(B) × P(R)

$= \frac{4}{7} \times \frac{3}{7} = \frac{12}{49}$ The AND rule

P(red and blue)

P(R) × P(B)

$= \frac{3}{7} \times \frac{4}{7} = \frac{12}{49}$ The AND rule

P(one of either colour)

$= \frac{12}{49} + \frac{12}{49} = \mathbf{\frac{24}{49}}$ The OR rule

Product Rule for Counting

In probability questions, the product rule for counting can be used to find the total number of combinations possible.

For example, if there are A ways of doing task 1 and B ways of doing task 2, there are $A \times B$ ways of doing both tasks.

SUMMARY

- **If two events are mutually exclusive, the probability of A or B happening is found by adding the probabilities.**

  P(A or B) = P(A) + P(B)

- **If two events are independent, the probability of A and B happening is found by multiplying the probabilities.**

  P(A and B) = P(A) × P(B)

- **These rules also work for more than two events.**
- **Remember that the probabilities on the branches leaving each point on a tree diagram should add up to 1.**

QUESTIONS

QUICK TEST

1. Sangeeta has a biased dice. The probability of getting a three is 0.4. She rolls the dice twice.

   a. Complete the tree diagram.

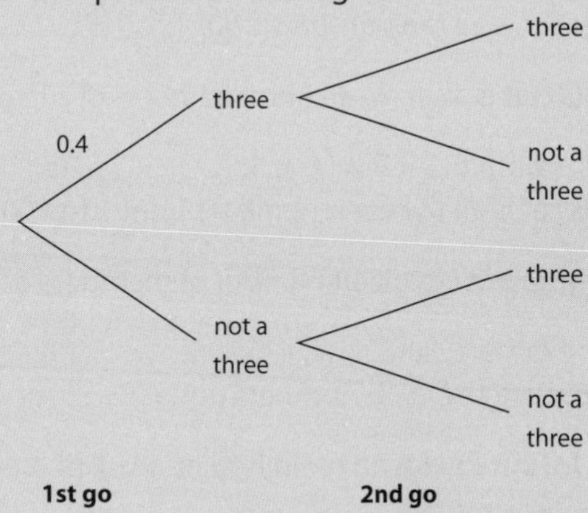

   b. Work out the probability that she gets:

      i. two threes

      ii. exactly one three

EXAM PRACTICE

1. A bag contains three red, four blue and two green beads. A bead is picked out of the bag at random and its colour noted. It is replaced in the bag. A second bead is picked out at random.

   Work out the probability that two different-coloured beads are chosen. [4 marks]

2. Mr Smith and Mrs Tate both go to the library. They can take out a fiction book or non-fiction book. The probability that Mr Smith takes out a fiction book is 0.8. The probability that Mrs Tate takes out a fiction book is 0.4. The events are independent.

   a. Calculate the probability that both Mr Smith and Mrs Tate take out a fiction book. [2 marks]

   b. Calculate the probability that one of each type of book is taken out. [3 marks]

Sets and Venn Diagrams

Set Notation

- A **set** is a collection of objects, which are called the elements or members of the set.

  For example: Set A = {1, 2, 3, …………..100}
  This set is all the numbers from 1 to 100.

- A **subset** is a set made from members of a larger set.

  For example: Set B = {2, 4, 6, …………..100}
  This set is all the even numbers from 1 to 100.

- A **finite set** is a given number of members of a set.

  For example: Set C = {2, 4, 6, …………..20}
  These are the even numbers up to 20.

- An **infinite set** is an unlimited number of members of a set.

  For example: Set D = {2, 4, 6, …………..}
  These are all the even numbers.

- An **empty** or **null set** Ø is a set that contains no members.

- A **universal set** is a set that contains all possible members.

Union of Sets (∪)

Set A ∪ B contains members belonging to A or B or both.

Example
If A = {2, 3, 4} and B = {4, 5, 6, 7}, then A ∪ B = {2, 3, 4, 5, 6, 7}.

Intersection of Sets (∩)

Set A ∩ B contains members belonging to both A and B.

Example
If A = {2, 3, 4} and B = {1, 3, 5}, then A ∩ B = {3} since this is the only number in both sets.

If A and B have no members in common, then A ∩ B = Ø, i.e. the empty set.

Venn Diagrams

Relationships between sets can be shown on a **Venn diagram**. The universal set contains all the elements being discussed and is shown as a rectangle.

Examples

1. A + A′ = ξ

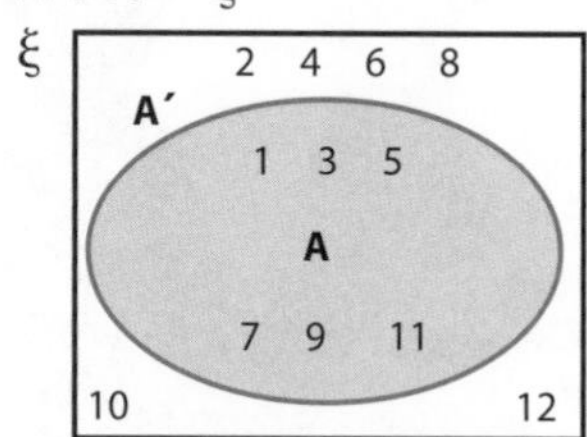

A = {1, 3, 5, 7, 9, 11}

A′ = {2, 4, 6, 8, 10, 12} ← A′ is the set of numbers **not** in set A.

ξ = {1, 2, 3, …….12}

2. A ∪ B

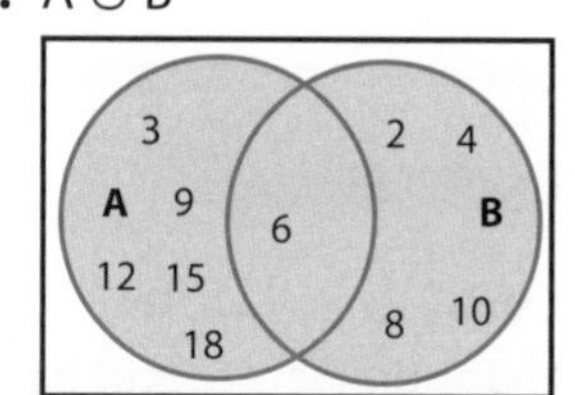

A = {3, 6, 9, 12, 15, 18}

B = {2, 4, 6, 8, 10}

A ∪ B = {2, 3, 4, 6, 8, 9, 10, 12, 15, 18}

3. A ∩ B

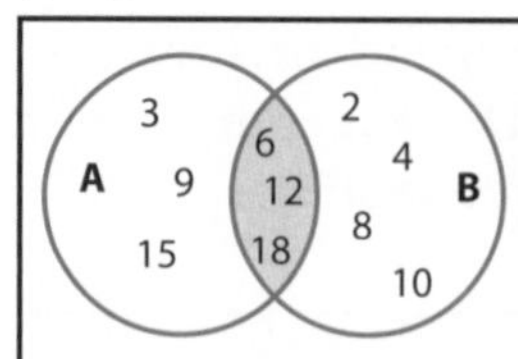

A = {3, 6, 9, 12, 15, 18}
These are multiples of 3.

B = {2, 4, 6, 8, 10, 12, 18}
These are multiples of 2.

A ∩ B = {6, 12, 18}
These are multiples of 6.

Using Venn Diagrams to Solve Probability Questions

Venn diagrams can be used to help solve probability questions.

Example

Out of 40 students, 14 are taking French and 29 are taking German.

a. If five students are in both classes, how many students are in neither class?

b. How many are in at least one class?

c. What is the probability that a randomly-chosen student is only taking German?

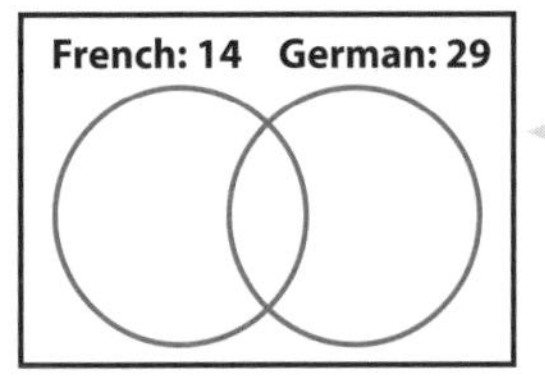

First draw the universal set for the 40 students, with two overlapping circles labelled with the total for each.

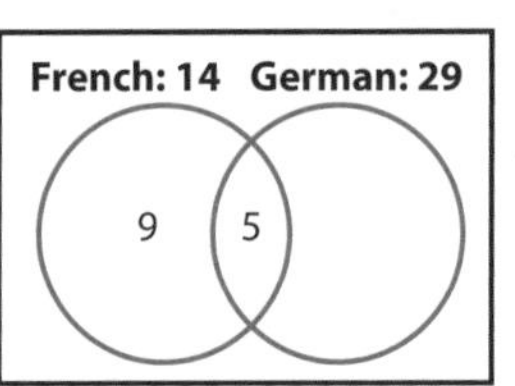

Since five students are taking both classes, put "5" in the overlap. Five of the 14 French students have been accounted for, leaving nine students taking French but not German, so put "9" in the "French only" part of the "French" circle.

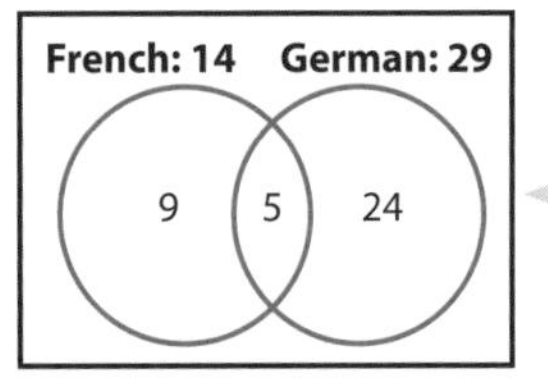

Five of the 29 German students have been accounted for, leaving 24 students taking German but not French, so put "24" in the "German only" part of the "German" circle.

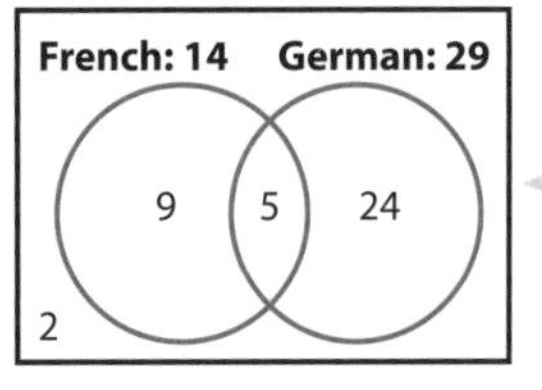

This tells you that a total of 9 + 5 + 24 = 38 students are taking either French or German (or both). This leaves two students unaccounted for, so they must be the ones taking neither class.

From this Venn diagram, the answers are:

a. **Two** students are in neither class.

b. There are **38** students in at least one of the classes.

c. There is a $\frac{\mathbf{24}}{\mathbf{40}} = \mathbf{0.6} = \mathbf{60\%}$ probability that a randomly-chosen student in this group is taking only German.

SUMMARY

- **Set A ∪ B contains members belonging to A or B or both.**
- **Set A ∩ B contains members belonging to both A and B.**
- **If A and B have no members in common, then A ∩ B = Ø, i.e. the empty set**
- **An empty or null set Ø contains no members.**
- **A universal set contains all possible members.**
- **Sets can be shown in a Venn diagram.**

QUESTIONS

QUICK TEST

1. If C = {2, 3, 4, 5, 6, 7} and D = {6, 7, 8, 9} then write down:

a. C ∪ D **b.** C ∩ D

EXAM PRACTICE

1. Tim asked 50 people which type of chocolate they liked from plain (P), milk (M) and white (W).
All 50 people liked at least one of the types.
19 people liked all three flavours.
16 people liked plain and milk chocolate but did not like white chocolate.
21 people liked milk and white chocolate.
24 people liked plain and white chocolate.
40 people liked milk chocolate.
1 person liked only white chocolate.

a. Draw a Venn diagram to represent this information. [3 marks]

Tim chose at random one of the 50 people.

b. Work out the probability that the person liked plain chocolate. [2 marks]

c. Given that the person selected at random liked plain chocolate, find the probability that this person liked exactly one other type of chocolate. [2 marks]

Statistical Diagrams

Diagrams to Compare Data

Pictograms, pie charts and vertical line graphs can be used to display data.

Dual bar charts can be used to compare data. In a dual bar chart, two (or more) bars are drawn side by side.

Example

Thomas has carried out a survey on some students' favourite sport. Here are his results.

Favourite sport	No. of boys	No. of girls
Swimming	6	8
Football	15	3
Hockey	5	10
Tennis	9	14

Draw a dual bar chart of these results.

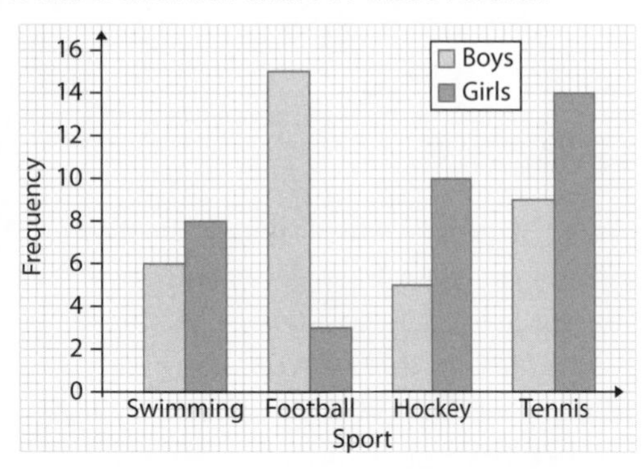

A compound bar chart can also compare two or more sets of data, e.g. for Thomas's results:

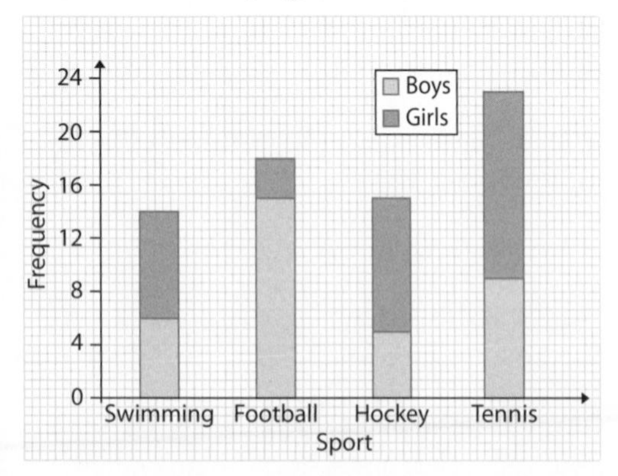

Both bar charts show that considerably more boys than girls like football and considerably more girls than boys like tennis.

Pie Charts

Pie charts are circles split up into sectors. Each sector represents a certain number of items.

Drawing a Pie Chart

When drawing a pie chart, you need to:

1. Calculate the angles:

- Find the total for the items listed.
- Work out how many degrees one item represents.
- Work out the degrees for each category.

2. Draw the pie chart accurately:

- You are only allowed to be at most 2° out!
- Label the sectors.

Example

The favourite subjects of 24 students are shown in the table. Draw a pie chart for this information.

Subject	Frequency
Maths	9
English	4
Art	5
Geography	6
	24

24 students = 360°

$$1 \text{ student} = \frac{360°}{24} = 15°$$

Multiply each frequency by 15° to find the angles. Now draw the pie chart carefully.

The angles must be accurate and each sector must be labelled.

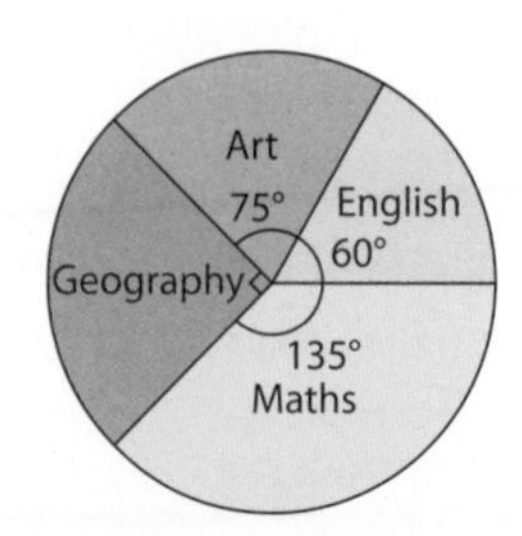

Interpreting a Pie Chart

When interpreting a pie chart, you need to carefully measure the angles if they are not given.

Example

The pie chart shows the favourite sports of 18 students. How many students like:

a. tennis? **b.** football? **c.** hockey?

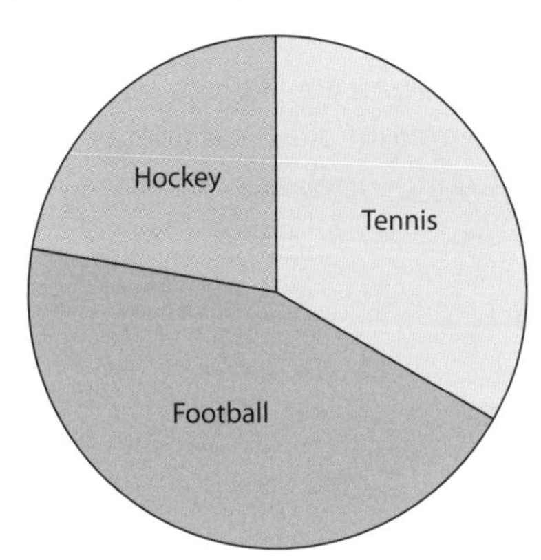

$360° = 18$ students so

1 student $= 20°$

Therefore:

Tennis (120°) $= \frac{120}{20} =$ **6 students**

Football (160°) $= \frac{160}{20} =$ **8 students**

Hockey (80°) $= \frac{80}{20} =$ **4 students**

SUMMARY

- **A dual bar chart has two or more bars drawn side by side. It can be used to compare data.**
- **When drawing a pie chart, you need to work out how many degrees one item represents and then multiply by the number of items in each category to find the angles.**
- **Always draw the pie chart accurately and label the sectors.**
- **When interpreting a pie chart, you need to work out what 1° or 1 item is worth and then use this unitary amount to find the number of items in each category.**

QUESTIONS

QUICK TEST

1. Audrey owns a flower shop. She has produced a bar chart to compare the percentages of types of flowers sold over the last two years.

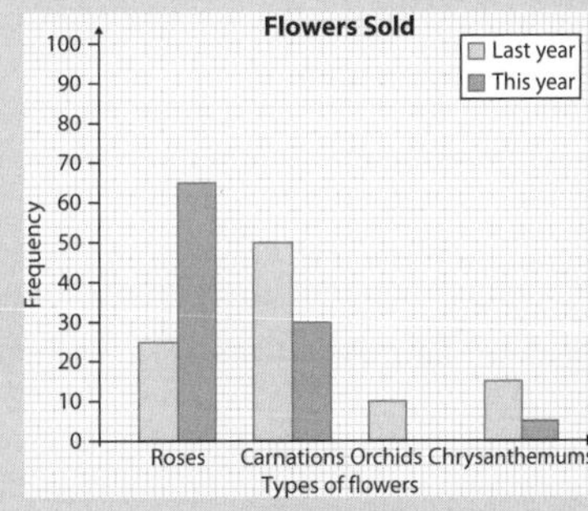

a. Write down the most popular flower sold:
 i. this year **ii.** last year

b. Work out the percentage of chrysanthemums sold:
 i. this year **ii.** last year

2. Draw a pie chart for this set of data.

Favourite colour	Frequency
Blue	15
Red	9
Black	5
Green	7

EXAM PRACTICE

1. A local authority wishes to build a new school and a new retirement home. It looks at the age profile of two towns:

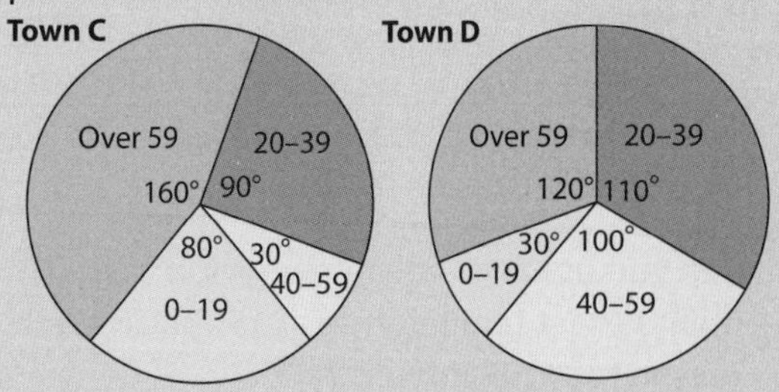

Town C has 75 40–59 year olds.
Town D has 500 40–59 year olds.

Using the information in the pie charts, state with reasons and calculations which town should have the school and which town should have the retirement home. [4 marks]

Scatter Diagrams and Time Series

Time Series

A **time-series graph** allows you to see how data is changing over time. Time-series graphs are used in hospitals for recording temperature over a period of time, the Retail Price Index, share prices, etc. Time is always plotted on the horizontal axis.

Example

Colin is in hospital. His temperature is taken every two hours and recorded in the table.

Time	Temperature (°C)
0600	38.6
0800	38.2
1000	38.8
1200	38.8
1400	38.0
1600	37.6
1800	37.6
2000	37.2

a. Show Colin's temperature on a time-series graph.

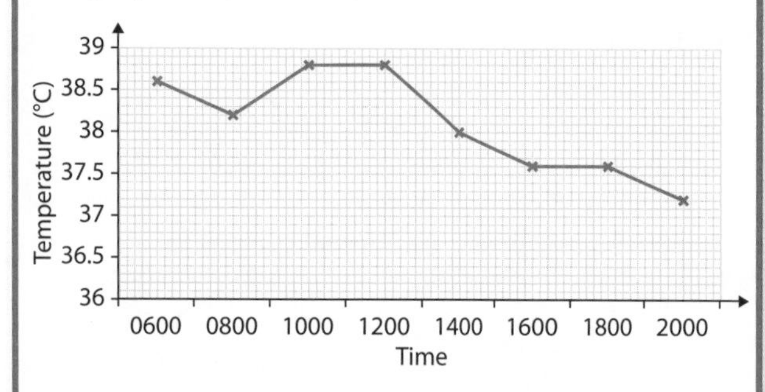

b. Can you use this time-series graph to find Colin's exact temperature at 1500? Give a reason for your answer.

You cannot use the graph to find the exact temperature at 1500. The values in between the plotted points have no meaning.

Scatter Diagrams and Correlation

Scatter diagrams are used to show two sets of data at the same time. They are important because they show the **correlation** (connection) between the sets.

Positive Correlation

Both variables are increasing. For example, the taller you are, the more you are likely to weigh. If the points are nearly in a straight diagonal line, as below, there is said to be a strong correlation.

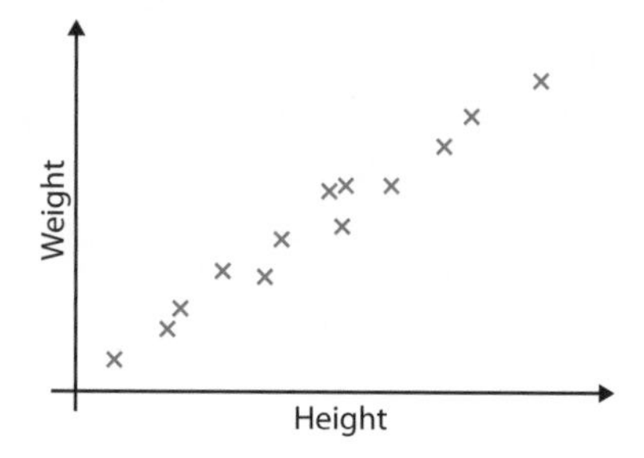

Negative Correlation

As one variable increases, the other decreases. For example, as the temperature increases, the sales of woollen hats are likely to decrease.

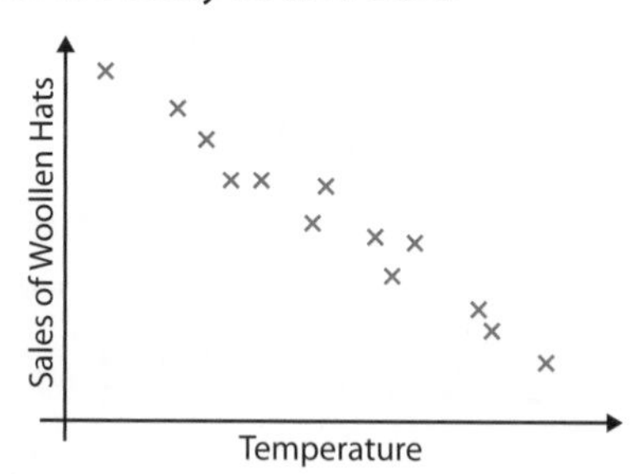

Zero Correlation

Little or no linear relationship between the variables. For example, there is no connection between your height and your mathematical ability.

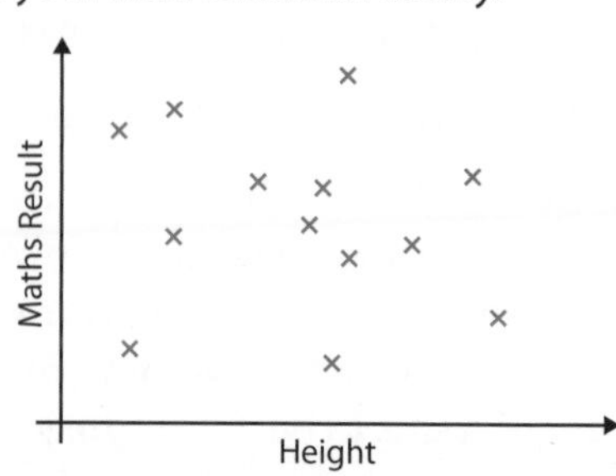

Line of Best Fit

The **line of best fit** goes as close as possible to all the points. There is roughly an equal number of points above the line and below it.

The scatter diagram below shows the Science and Maths percentages scored by some students.

- The line of best fit goes in the direction of the data.
- The line of best fit can be used to estimate results.
- We can estimate that a student with a Science percentage of 30 would get a Maths percentage of about 18.
- We can estimate that a student with a Maths percentage of 50 would get a Science percentage of about 54.

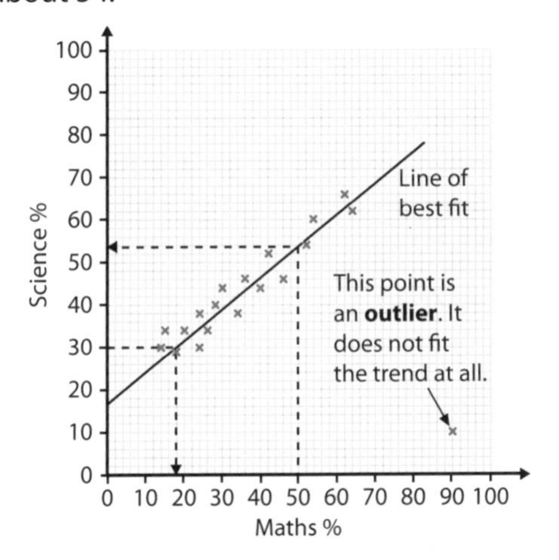

SUMMARY

- **Time-series line graphs show trends in the data. The points are plotted and joined with straight lines. Some form of time measurement is on the horizontal axis.**
- **Positive correlation: both variables are increasing.**
- **Negative correlation: as one variable increases, the other decreases.**
- **Zero correlation: little or no linear correlation between variables.**
- **A line of best fit should be as close as possible to all the points. It is in the direction of the data.**

QUESTIONS

QUICK TEST

1. The table shows shop sales of children's bikes.

Month	Number of bikes
March (Mar)	3
April (Apr)	6
May (May)	10
June (Jun)	12
July (Jul)	14
August (Aug)	10
September (Sep)	5

a. Draw a time-series graph for this data, on a scale similar to the one below.

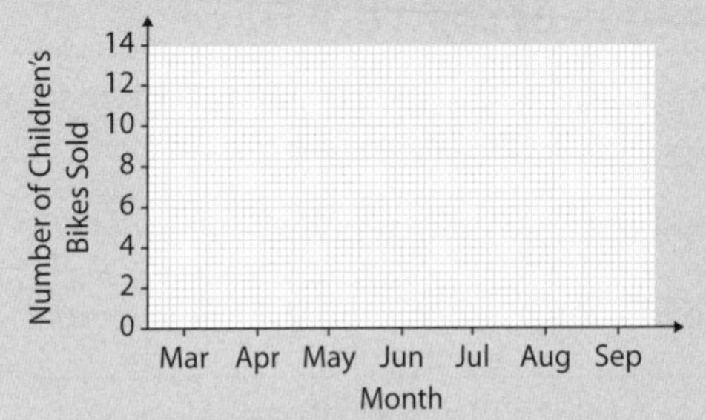

b. Describe the sales pattern shown by the graph.

EXAM PRACTICE

1. The scatter diagram shows car ages and values.

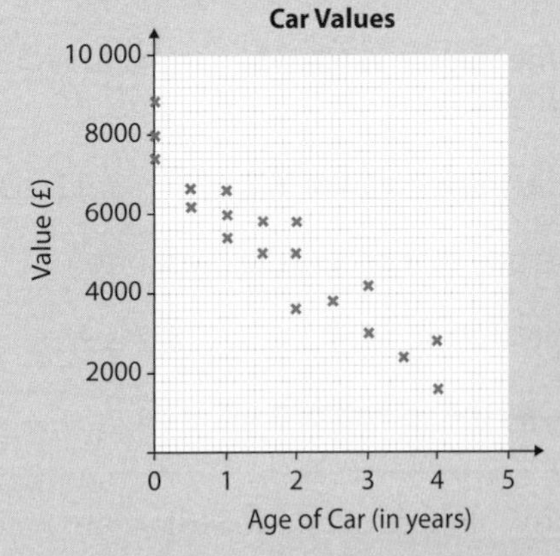

a. Draw a line of best fit on the diagram. [1 mark]

b. Use your line of best fit to estimate the age of a car when its value is £5000. [2 marks]

c. Use your line of best fit to estimate the value of a $3\frac{1}{2}$ year-old car. [2 marks]

d. Why can the graph not be used to predict the value of a 10-year-old car? [1 mark]

Averages 1

Sampling

Sampling is an efficient way of collecting information about a **population**. It is important that the sample is representative of the population and does not contain **bias**. The bigger the sample, the more accurate the results will be.

In a **random sample**, each member of the population will have an equally likely chance of being chosen. In **stratified sampling**, the population is divided into groups (strata) that have something in common. The same proportion of the group is used for each sample.

Averages of Discrete Data

Averages are used to give an idea of a 'typical' value for a set of data. **Discrete data** has an exact value.

You should know these types of averages:

1. **Mean** – the most commonly used average:

$$\text{Mean} = \frac{\text{sum of a set of values}}{\text{the number of values used}}$$

For example, the mean of 1, 2, 3, 3, 1 is $\frac{1+2+3+3+1}{5} = 2$

2. **Median** – the middle value when the values are put in order of size.

For example, the median of 2, 2, 3, 3, 7, 9, 11 is 3

3. **Mode** – the one that occurs the most often.

For example, the mode of 2, 2, 2, 3, 5, 7 is 2

The **range** shows the spread of a set of data.

Range = highest value – lowest value

For example, the range of 1, 2, 3, 4, 7, 10 is 10 – 1 = 9

Finding Averages from a Frequency Table

When the information is in a frequency table, finding the averages is a little more difficult.

To find the mean of a frequency table, we use:

$$\text{mean } (\bar{x}) = \frac{\Sigma fx}{\Sigma f}$$

Σ means the sum of
f represents the frequency
$\bar{x}$ represents the mean

Example

The table shows the shoe sizes of a group of students. Find the mean, range, mode and median.

Shoe size (x)	3	4	5	6	7
Frequency (f)	5	18	22	15	5

Mean shoe size

$$\text{Mean} = (\bar{x}) = \frac{\Sigma fx}{\Sigma f}$$

- Multiply the frequency, f, by x $= (3 \times 5) + (4 \times 18) + (5 \times 22) + (6 \times 15) + (7 \times 5)$
- Add up the frequency $= 5 + 18 + 22 + 15 + 5$
- Divide the sum of fx by the sum of f:

$$\frac{(3 \times 5) + (4 \times 18) + (5 \times 22) + (6 \times 15) + (7 \times 5)}{5 + 18 + 22 + 15 + 5}$$

$= \frac{322}{65} =$ **4.95** (2 d.p.)

Range

Range = highest value – lowest value

Range = 7 – 3 = **4**

Mode

Modal shoe size = **5** (highest frequency)

Median

Median shoe size: $\frac{\Sigma f + 1}{2}$ tells you how many places along the list to go.

33rd person has size 5, so median = **5**

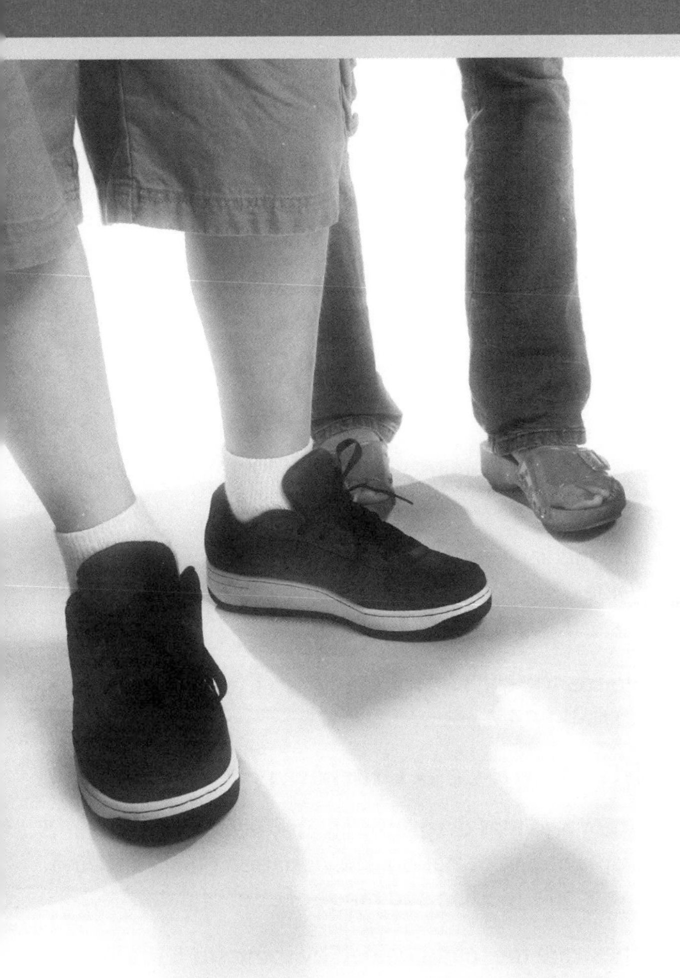

SUMMARY

- **Sampling is an efficient way of collecting information about a population.**
- **In a random sample, each member of the population will have an equally likely chance of being chosen.**
- **For discrete data:**
  - **Mean = $\frac{\text{sum of a set of values}}{\text{the number of values used}}$**
  - **Median = middle value when values are put in order of size**
  - **Mode = the value that occurs most often**
- **Range = highest value – lowest value**
- **Find the mean from a frequency table:**

$$\text{mean } (\bar{x}) = \frac{\Sigma fx}{\Sigma f}$$

Σ means the sum of
f represents the frequency
$\bar{x}$ represents the mean

QUESTIONS

QUICK TEST

1. Find the mean and range of each set of data:

   a. 2, 7, 9, 3, 6, 4, 5, 2 **b.** 7, 9, 11, 15, 2, 1, 6, 12, 19, 13

2. Find the median and mode of each set of data:

   a. 2, 7, 1, 4, 2, 2, 3, 7, 9, 2 **b.** 6, 9, 11, 11, 13, 6, 9, 6, 6, 4, 1

EXAM PRACTICE

1. Reece created this table for the number of minutes (to the nearest minute) it took some students to complete a Maths problem.

   Calculate:

   a. the mean [3 marks]

   b. the median [1 mark]

   c. the mode [1 mark]

Number of minutes to solve problem	5	6	7	8	9	10
Frequency	4	7	10	4	3	1

   d. Rupinder says: 'The mean number of minutes to solve the problem is 12 minutes'.

   Explain why Rupinder cannot be correct. [1 mark]

Averages 2

Averages of Continuous Data

When data is grouped into **class intervals**, the exact values are not known.

You estimate the **mean** by using the midpoints of the class intervals.

Add in two extra columns – one for the midpoint and one for fx.

Weight (W kg)	Frequency (f)	Midpoint (x)	fx
$30 \leqslant W < 35$	6	32.5	195
$35 \leqslant W < 40$	14	37.5	525
$40 \leqslant W < 45$	22	42.5	935
$45 \leqslant W < 50$	18	47.5	855
	60		**2510**

Σf Σfx

For continuous data:

$$\bar{x} = \frac{\Sigma fx}{\Sigma f}$$

Σ means the sum of
f represents the frequency
$\bar{x}$ represents the mean
x represents the midpoint of the class interval

$$\bar{x} = \frac{\Sigma fx}{\Sigma f}$$

$$\bar{x} = \frac{2510}{60}$$

$$\bar{x} = 41.8\dot{3}$$

Modal class is $40 \leqslant W < 45$

This class interval has the highest frequency.

To find the class interval containing the **median**, first find the position of the median:

$$= \frac{\Sigma f + 1}{2}$$

$$= \frac{60 + 1}{2} = 30.5$$

The median lies halfway between the 30th and 31st values. The 30th and 31st values are in the class interval $40 \leqslant W < 45$.

Hence, the class interval in which the median lies is $40 \leqslant W < 45$.

Stem and Leaf Diagrams

Stem and leaf diagrams are useful for recording and displaying information. They can also be used to find the mode, median and range of a set of data.

These are the marks gained by some students in a Maths test:

52 45 63 67

75 57 68 67

60 59 67

In an ordered stem and leaf diagram, it would look like this:

4	5					
5	2	7	9			
6	0	(3)	7	7	7	8
7	5					

Median (circled 3)

Key 6 | 3 = 63 marks

The median is the sixth value = 63 marks

The mode is 67 marks.

The range (highest value – lowest value) is 75 – 45 = 30 marks

If the same students sat a second Maths test, their results could be put into a back-to-back stem and leaf diagram. These are very useful when comparing two sets of data.

Test 2		Test 1
9 6 2	4	5
9 7 5 1 1	5	2 7 9
3 1 0	6	0 3 7 7 7 8
	7	5

Key Test 1 5 | 2 = 52 marks
Test 2 1 | 5 = 51 marks

Comparing the data in the back-to-back stem and leaf diagram, you can say that:

'In test 2, the median score of 55 marks is lower than the median score in test 1 of 63 marks. The range of the scores in test 2 is 21 marks, which is lower than the range of the scores in test 1 of 30 marks. So on average, the students did better in test 1 but their scores were more variable than in test 2.'

SUMMARY

- **For continuous data:**
  - **Estimate the mean using:**

$$\bar{x} = \frac{\Sigma fx}{\Sigma f}$$

Σ means the sum of
f represents the frequency
$\bar{x}$ represents the mean
x represents the midpoint of the class interval

  - **Modal class is the class interval with the highest frequency.**
- **Use stem and leaf diagrams to record and display information. Do not forget to order the leaves and write a key.**
- **Back-to-back stem and leaf diagrams can be used to compare two sets of data.**

QUESTIONS

QUICK TEST

1. The heights, h cm, of some students are shown in the table.

Height (h cm)	Frequency	Midpoint	fx
$140 \leqslant h < 145$	4		
$145 \leqslant h < 150$	9		
$150 \leqslant h < 155$	15		
$155 \leqslant h < 160$	6		

Calculate an estimate for the mean of this data.

2. a. Draw an accurate stem and leaf diagram of this data.

27 28 36 42 50 18
25 31 39 25 49 31
33 27 37 25 47 40
7 31 26 36 9 42

b. What is the median of this data?

EXAM PRACTICE

1. The table shows information about the number of hours that 50 children watched television for last week.

Work out an estimate for the mean number of hours the children watched television.

[4 marks]

Number of hours (h)	Frequency
$0 \leqslant h < 2$	3
$2 \leqslant h < 4$	6
$4 \leqslant h < 6$	22
$6 \leqslant h < 8$	13
$8 \leqslant h < 10$	6

Answers

You are encouraged to show your working out, as you may be awarded marks for method in your exam even if your final answer is wrong. Full marks can be awarded where a correct answer is given without working being shown, but if a question asks for working out, you must show it to gain full marks. If you use a correct method that is not shown in the mark scheme below, you would still gain full credit for it.

N.B. In the answers to Exam Practice Questions, **[1]** *indicates where individual marks may be awarded for correct workings and method.*

Day 1

pages 4–5

Prime Factors, HCF and LCM

QUICK TEST

1. **a.** $50 = 2 \times 5 \times 5$ **or** 2×5^2

   b. $360 = 2 \times 2 \times 2 \times 3 \times 3 \times 5$ **or** $2^3 \times 3^2 \times 5$

   c. $16 = 2 \times 2 \times 2 \times 2$ **or** 2^4

2. **a.** False

   b. True

   c. True

   d. False

EXAM PRACTICE

1. $120 = 2 \times 2 \times 2 \times 3 \times 5$ **[1]**

   $42 = 2 \times 3 \times 7$ **[1]**

   HCF = 6 **[1]**

2. LCM of 20 and 14

   $20 = 2 \times 2 \times 5$

   $14 = 2 \times 7$

   $\text{LCM} = 2 \times 2 \times 5 \times 7$ **[1]**

   LCM = 140

   Both buses will leave at the same time 140 minutes later, i.e. 2 hours and 20 minutes. **[1]**

   Time they leave together = 12.20 pm **[1]**

   You could also list both the times of the buses from Hatfield and St Albans and find the time that is the same in both lists.

pages 6–7

Fractions and Decimals

QUICK TEST

1. **a.** $\frac{13}{15}$

   b. $2\frac{11}{21}$

   c. $\frac{10}{63}$

   d. $\frac{81}{242}$

2. **a.** $5\frac{7}{10}$

   b. $1\frac{53}{90}$

   c. $\frac{32}{75}$

   d. 14

EXAM PRACTICE

1. Let x = number of pages in the magazine

   $\frac{3}{7}x = 12$

   $x = (12 \times 7) \div 3$ **[1]**

   $x = 28$ pages **[1]**

2. $\frac{3}{4} \times 60 = 45$ minutes **[1]**

   $\frac{2}{3} \times 60 = 40$ minutes **[1]**

   $45 + 40 = 1$ hour 25 minutes

   Total time = 3 hours 25 minutes **[1]**

   Or

   $\frac{3}{4} + \frac{2}{3} =$

   $\frac{9}{12} + \frac{8}{12} = \frac{17}{12} = 1\frac{5}{12}$ **[1]**

   $\frac{5}{12} \times 60 = 25$ minutes **[1]**

   Total time = 3 hours 25 minutes **[1]**

3. $\frac{3}{4}$, 0.4, 0.85, $\frac{8}{10}$, $\frac{675}{1000}$

   $\frac{3}{4} = 0.75$

   $\frac{8}{10} = 0.8$

   $\frac{675}{1000} = 0.675$ **[1]**

   In order:

   0.4, $\frac{675}{1000}$, $\frac{3}{4}$, $\frac{8}{10}$, 0.85 **[1]**

pages 8–9

Rounding and Estimating

QUICK TEST

1. 3700
2. a. True

   b. False

   c. True

   d. False

EXAM PRACTICE

1. $\frac{300 \times 3}{0.05}$ **[1]** $= \frac{900}{0.05}$ **[1]** $= \frac{90\,000}{5} = 18\,000$ **[1]**
2. 27.5 grams

pages 10–11

Indices

QUICK TEST

1. a. 6^8

   b. 12^{13}

   c. 5^6

   d. 4^2

2. a. $6b^{10}$

   b. $2b^{-16}$

   c. $9b^8$

   d. $\frac{1}{25x^4y^6}$ **or** $\frac{1}{25}x^{-4}y^{-6}$

EXAM PRACTICE

1. a. 1

   b. $\frac{1}{7^2} = \frac{1}{49}$

   c. $\sqrt[3]{64} \times \sqrt{144}$ **[1]**

   $= 4 \times 12$

   $= 48$ **[1]**

   d. $\frac{1}{27^{\frac{2}{3}}} = \frac{1}{(\sqrt[3]{27})^2}$ **[1]**

   $= \frac{1}{9}$ **[1]**

2. a. $\frac{x^{11}}{x^{15}}$ **[1]** $= x^{-4}$ **[1]**

   or $\frac{1}{x^4}$

   b. $\frac{12x^6}{2x^3}$ **[1]** $= 6x^3$ **[1]**

pages 12–13

Standard Index Form

QUICK TEST

1. a. 6.4×10^4

   b. 4.6×10^{-4}

2. a. 1.2×10^{11}

   b. 2×10^{-1}

3. a. 1.4375×10^{18}

   b. 5.48×10^{19}

EXAM PRACTICE

1. a. 4×10^7

   b. 0.000 06

2. $2 \times 10^{-23} \times 7 \times 10^{16}$ **[1]**

   $= 14 \times 10^{-7}$ **[1]**

   $= 1.4 \times 10^{-6}$ g **[1]**

   Make sure that you check that your final answer is written in standard form.

pages 14–15

Formulae and Expressions 1

QUICK TEST

1. a. $4a$

   b. $8a + b$

   c. $8a - 7b$

   d. $15xy$

   e. $4a^2 - 8b^2$

   f. $2xy + 2xy^2$

EXAM PRACTICE

1. a. $4bc - 2ab$ **[1 for $4bc$; 1 for $-2ab$]**

   b. d^4

   c. $15mn$

2. $T = 7x + 0.98y$ **[1 for $T =$; 1 for $7x$; 1 for $0.98y$]**

   or $T = 700x + 98y$ **[1 for $T =$; 1 for $700x$; 1 for $98y$]**

3. $C = 2x + 4y$ **[1 for $C =$; 1 for $2x$; 1 for $4y$]**

Answers

pages 16–17

Formulae and Expressions 2

QUICK TEST

1. **a.** $-\frac{31}{5} = -6\frac{1}{5}$

   b. 4.36 **or** $\frac{109}{25}$

   c. 9

2. $u = \pm\sqrt{v^2 - 2as}$

EXAM PRACTICE

1. $5a - b = 3p + 2b$

   $5a - b - 2b = 3p$ **[1]**

   $5a - 3b = 3p$ **[1]**

   $p = \frac{5a - 3b}{3}$ **[1]**

2. $b = \frac{m}{h^2}$ $m = 84.5$ $h = 1.79$ m

   $b = \frac{84.5}{1.79^2}$ **[1]**

   $b = 26.37$ **[1]**

   Yes, Dan would be classed as overweight. **[1]**

Day 2

pages 18–19

Brackets and Factorisation

QUICK TEST

1. **a.** $x^2 + x - 6$

   b. $4x^2 - 12x$

   c. $x^2 - 6x + 9$

2. **a.** $6x(2y - x)$

   b. $3ab(a + 2b)$

   c. $(x + 2)(x + 2) = (x + 2)^2$

   d. $(x + 1)(x - 5)$

   e. $(x + 10)(x - 10)$

EXAM PRACTICE

1. **a.** $3t^2 - 4t$

   b. $4(2x - 1) - 2(x - 4) = 8x - 4 - 2x + 8$ **[1]**

   $= 6x + 4$ **[1]**

2. **a.** $y(y + 1)$

   b. $5pq(p - 2q)$ **[1 for $5pq$; 1 for $(p - 2q)$]**

   c. $(a + b)(a + b + 4)$

   d. $(x - 2)(x - 3)$ **[1 for identifying 2 and 3 as factors]**

3. $(a + b)^2 - 2b(a + b)$

   $= (a + b)(a + b) - 2b(a + b)$

   $= (a + b)(a + b) - 2ab - 2b^2$

   $= a^2 + 2ab + b^2 - 2ab - 2b^2$ **[1]**

   $= a^2 - b^2$ **[1]**

   $= (a - b)(a + b)$ **[1]**

Always check that the final algebraic expression is in its simplest form – if not, then simplify by factorising.

pages 20–21

Equations 1

QUICK TEST

1. $x = 8$
2. $x = -5$
3. $x = -\frac{1}{2}$
4. $x = -3.25$
5. $x = -\frac{1}{2}$
6. $x = 17$

EXAM PRACTICE

1. **a.** $5x - 3 = 9$

   $5x = 9 + 3$ **[1]**

   $5x = 12$

   $x = \frac{12}{5}$ **[1]**

   $x = 2.4$

   b. $7x + 4 = 3x - 6$

   $7x + 4 - 3x = -6$ **[1]**

   $4x = -6 - 4$

   $4x = -10$ **[1]**

   $x = -\frac{10}{4}$

   $x = -2.5$ **[1]**

c. $3(4y - 1) = 21$

$12y - 3 = 21$ **[1]**

$12y = 21 + 3$ **[1]**

$12y = 24$

$y = 2$ **[1]**

2. a. $5 - 2x = 3(x + 2)$

$5 - 2x = 3x + 6$ **[1]**

$5 = 3x + 6 + 2x$ **[1]**

$5 - 6 = 5x$

$x = -\frac{1}{5}$ **[1]**

b. $\frac{3x - 1}{3} = 4 + 2x$

$3x - 1 = 3(4 + 2x)$ **[1]**

$3x - 1 = 12 + 6x$

$-1 - 12 = 6x - 3x$ **[1]**

$-13 = 3x$

$x = -\frac{13}{3}$ **[1] or** $x = -4\frac{1}{3}$

3. Joe forgot to multiply the 3 and the –6 together. When he multiplied out the bracket, it should say: $3x - 18 = 42$

pages 22–23

Equations 2

QUICK TEST

1. $x = 5.5$ cm
Shortest length:
$2x - 5 = 6$ cm

2. a. $x = 0, x = 7$

b. $x = -5, x = -3$

c. $x = 3, x = 2$

EXAM PRACTICE

1. $x + 30 + x + 50 + x + 10 + 2x = 360$ **[1]**

$5x + 90 = 360$

$5x = 360 - 90$ **[1]**

$5x = 270$

$x = 54$ **[1]**

Smallest angle = 64° **[1]**

2. $\frac{x - 2}{4} + \frac{x + 3}{2} = \frac{9}{4}$

$(x - 2) + 2(x + 3) = 9$ **[1]**

$x - 2 + 2x + 6 = 9$ **[1]**

$3x + 4 = 9$

$3x = 5$

$x = \frac{5}{3}$ **[1]**

pages 24–25

Simultaneous Linear Equations

QUICK TEST

1. $a = 4$ $b = -4.5$ **2.** $x = 2$ $y = 4$

EXAM PRACTICE

1. $5a - 2b = 19$ Equation 1

$3a + 4b = 1$ Equation 2

$15a - 6b = 57$ Equation 3 **[1]**

$15a + 20b = 5$ Equation 4

Equation 4 – Equation 3

$26b = -52$ so $b = -2$ **[1]**

Substitute $b = -2$ into Equation 1

$5a - (2 \times -2) = 19$ **[1]**

$5a + 4 = 19$

$5a = 19 - 4$

$5a = 15$, so $a = 3$ **[1]**

$a = 3, b = -2$

The equations could have also been solved by eliminating b.

2. $5x + 4y = 22$ **[1]** Equation 1

$3x + 5y = 21$ Equation 2

$15x + 12y = 66$ **[1]**

$15x + 25y = 105$

$13y = 39$, so $y = 3$ **[1]**

$5x + (4 \times 3) = 22$ **[1]**

$5x = 10$, so $x = 2$ **[1]**

Hats = £2 each and balloons = £3 each

Always check that your solutions work by substituting them back into Equation 2.

pages 26–27

Sequences

QUICK TEST

1. a. $4n + 1$

   b. $2 - n$

   c. $2n$

   d. $3n + 2$

   e. $5n - 1$

2. a. 2048, 8192

   b. Multiply the previous term by 4.

EXAM PRACTICE

1. 5 7 9 11
   The difference between the terms = 2 **[1]**
   $2 \times 1 = 2$, adjust by adding 3
   nth term = $2n + 3$ **[1]**

2. $2n^2 + 1 = 101$
   $2n^2 = 101 - 1$
   $2n^2 = 100$
   $n^2 = 50$ **[1]**
   Chloe is not correct. Since 50 is not a square number, 101 is not in the sequence. **[1]**

pages 28–29

Inequalities

QUICK TEST

1. a. $x < 2.2$

   b. $\frac{4}{3} \leqslant x < 3$

   c. $x > -\frac{9}{5}$

EXAM PRACTICE

1. −2, −1, 0, 1, 2, 3, 4

2. $4 + x > 7x - 8$
   $4 + 8 > 7x - x$ **[1]**
   $12 > 6x$
   $x < 2$ **[1]**

3. a. $\frac{3x + 5}{4} \leqslant 5$
   $3x + 5 \leqslant 20$ **[1]**
   $3x \leqslant 15$ **[1]**
   $x \leqslant 5$ **[1]**

Check that you still have the inequality sign in the answer and that you have not accidentally replaced it with an '=' sign.

b.

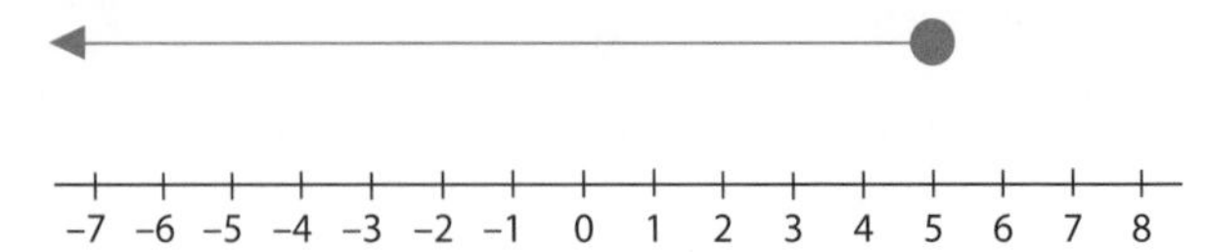

Day 3

pages 30–31

Straight-line Graphs

QUICK TEST

1. a.

x	−2	−1	0	1	2	3
y	−1	1	3	5	7	9

b.

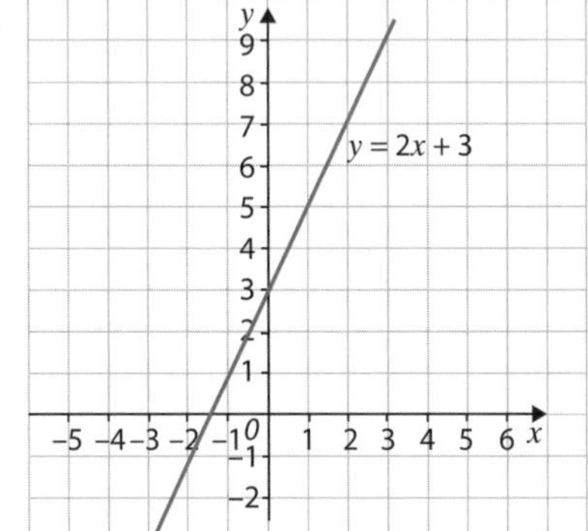

EXAM PRACTICE

1. a. Gradient of EF = $\frac{\text{change in } y}{\text{change in } x}$
   $= \frac{(6 - -4)}{(-2 - 3)}$ **[1]**

   Gradient of EF = $\frac{10}{-5} = -2$ **[1]**

   Goes through G (0, 3)

   Equation of L: $y = -2x + 3$ **[1]**

   b. Midpoint of EF $\left(\frac{-2 + 3}{2}, \frac{6 - 4}{2}\right)$ **[1]**
   Midpoint = $\left(\frac{1}{2}, 1\right)$ **[1]**

   c. Equation of line:
   $y = kx + 3$, where k can be any value, other than −2. **[1 for kx; 1 for +3]**

pages 32–33

Curved Graphs

QUICK TEST

1. Graph A: $y = \frac{3}{x}$
   Graph B: $y = 4x + 2$
   Graph C: $y = x^3 - 5$
   Graph D: $y = 2 - x^2$

EXAM PRACTICE

1. a.

x	–3	–2	–1	0	1	2	3
y	–28	–9	–2	–1	0	7	26

[2 if fully correct; 1 if one error]

b.

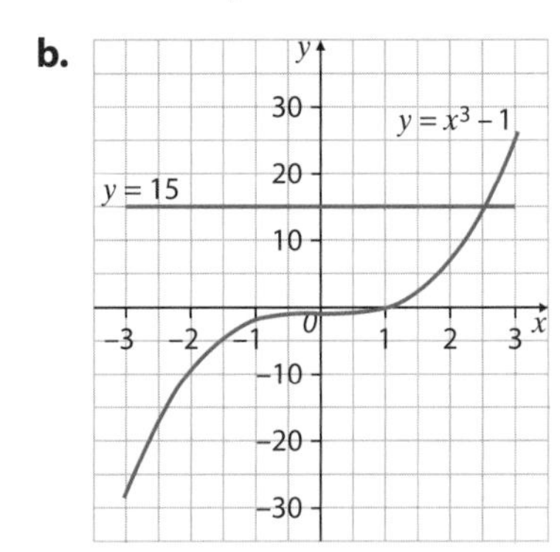

[1 for correct labels; 2 for correct points joined by a curve]

c. $x = 2.5$ **[1 for line at $y = 15$; 1 for $x = 2.5$]**

pages 34–35

Percentages

QUICK TEST

1. a. 6 kg
   b. £600
   c. £3
   d. 252 g
2. a. 32%
   b. 23%
   c. 75%
   d. 84%
3. £180
4. 80%

EXAM PRACTICE

1. Best TV Shop: $\frac{1}{5} \times 556 = £111.20$
   Cost = £556 – 111.20 = £444.80 **[1]**
   Drymons: 0.9 × 495 = £445.50 **[1]**
   Mark's Electricals: 385 × 1.2 = £462 **[1]**
   Since the television is the cheapest in the Best TV Shop, this is where Jonathan should purchase it. **[1]**
2. Taxable income: £10 600 (no tax paid)
   £48 500 – £10 600 = £37 900 to be taxed **[1]**
   $\frac{20}{100} \times 31\,786 = £6357.20$ **[1]**
   £37 900 – £31 786 = £6114 at 40% tax **[1]**
   $\frac{40}{100} \times 6114 = £2445.60$ **[1]**
   Total tax paid = £6357.20 + £2445.60
   = £8802.80 **[1]**

pages 36–37

Repeated Percentage Change

QUICK TEST

1. 51.7% (3 s.f.)
2. £121 856

EXAM PRACTICE

1. Savvy Saver: $\frac{2.5}{100} \times 3000 = £75$
   £75 × 2 = £150 **[1]**
   Money Grows: $1.025^2 \times 3000 = £3151.88$
   Interest earned = £151.88 **[1]**
   Shamil is not correct – he would earn £1.88 more with the Money Grows investment. **[1]**

The use of the multiplier in these questions enables you to answer the question in an efficient manner.

pages 38–39

Reverse Percentage Problems

QUICK TEST

1. a. £57.50
   b. £127
   c. £237.50
   d. £437.50

EXAM PRACTICE

1. 100% – 12% = 88%
   0.88 is the multiplier. **[1]**
   $0.88x = 220$

$x = \frac{220}{0.88}$ **[1]**

So x = £250 **[1]**

2. Yes, Joseph is correct **[1]** since $\frac{60}{0.85}$ = £70.59 **[2]**

Day 4

pages 40–41

Ratio and Proportion

QUICK TEST

1. £20 : £40 : £100
2. £35.28
3. 3 days

EXAM PRACTICE

1. There are many ways to work this out – this is one possible way:

   Work out the cost of 25 ml for each tube of toothpaste.
   50 ml = £1.24: 25 ml = 62p
   75 ml = £1.96: 25 ml = 65.$\dot{3}$p
   100 ml = £2.42: 25 ml = 60.5p **[2]**
   The 100 ml tube of toothpaste is the best value for money. **[1]**

2. £52 × \$1.49 = \$77.48 **[1] or**
   \$63 ÷ 1.49 = £42.28
   Cheaper in America by £9.72 (or by \$14.48) **[1]**

pages 42–43

Proportionality

QUICK TEST

1. **a.** $y = kx$

   b. $y = \frac{k}{x}$

EXAM PRACTICE

1. $a \propto x$

   $a = kx$ **[1]**

   $8 = 4k$

   $k = 2$ **[1]**

   $a = 2x$

   Constant of proportionality = 2

   $64 = 2x$ **[1]**

   $x = 32$ **[1]**

2. **a.** $y = \frac{k}{x}$ **[1]**

   $10 = \frac{k}{3}$

   $k = 30$ **[1]**

   $y = \frac{30}{2}$ **[1]**

   $y = 15$ **[1]**

   b. $6 = \frac{30}{x}$ **[1]**

   $x = 5$ **[1]**

pages 44–45

Measurement

QUICK TEST

1. 3.2 kg
2. 0330
3. 3 hours 26 minutes (to the nearest minute)

EXAM PRACTICE

1. Density = $\frac{\text{mass}}{\text{volume}}$

   Mass of solid A = 320 × 2.5 = 800 kg **[1]**

   Mass of solid B = 288 × 2.5 = 720 kg **[1]**

   Difference in mass = 800 – 720

   = 80 kg **[1]**

   A formula triangle can be useful when trying to remember the formulae for speed, density and pressure.

2. 5 miles ≈ 8 km

   Josie: 70 mph = 70 × $\frac{8}{5}$ **[1]**

   = 112 km/h **[1]**

   Jack: speed = 120 km/h

   Hence, Jack is travelling faster. **[1]**

pages 46–47

Interpreting Graphs

QUICK TEST

1. 48 km/h
2. 3 mph
3. A – graph 2 B – graph 3
   C – graph 1 D – graph 4

EXAM PRACTICE

1. a. 66.7 mph
   b. 1 hour
   c. 37.5 mph
   d. 0820
   e. 50 mph

pages 48–49

Similarity

QUICK TEST

1. a. 13.8 cm
   b. 4.5 cm

EXAM PRACTICE

1. No, Lucy is not correct. **[1]** The ratios of the corresponding lengths are not the same.

   $\frac{7}{3.5} = 2$ is not the same ratio as

   $\frac{3.7}{1.3} = 2.84\ldots$ **[1]**

2. $\frac{x}{12.4} = \frac{16.2}{9.7}$ **[1]**

   $x = \frac{16.2}{9.7} \times 12.4$ **[1]**

   $x = 20.7$ cm (3 s.f.) **[1]**

pages 50–51

2D and 3D Shapes

QUICK TEST

1. Hexagon
2. Joe is not correct because there is no 90° angle in the triangle.

EXAM PRACTICE

1. **[2]**

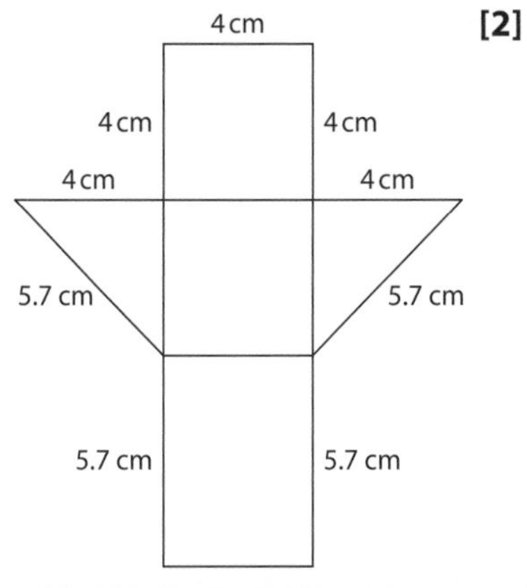

2. a. 5 b. 8

3. **[2]**

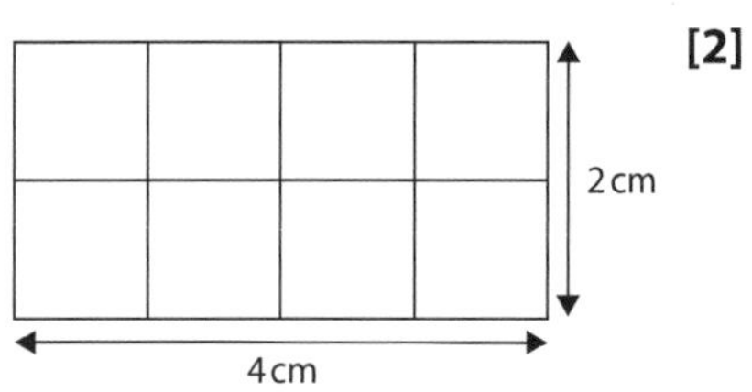

Day 5

pages 52–53

Constructions

QUICK TEST

1. Construct an equilateral triangle first. Bisect the 60° angle.

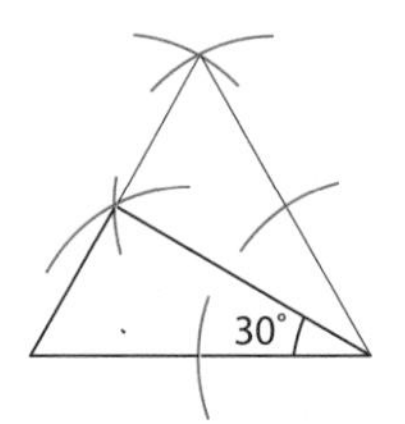

2. Perpendicular bisector of an 8 cm line should be drawn.

EXAM PRACTICE

1.

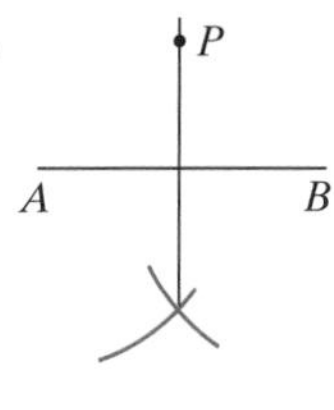

[2]

2.

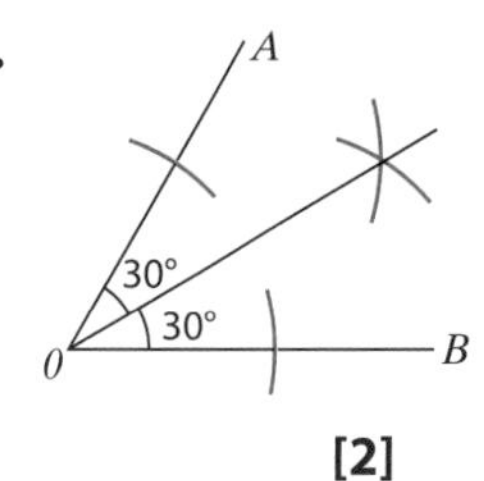

[2]

pages 54–55

Loci

QUICK TEST

1.

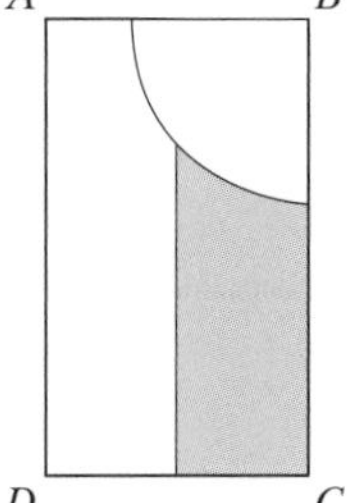

EXAM PRACTICE

1. a. b.

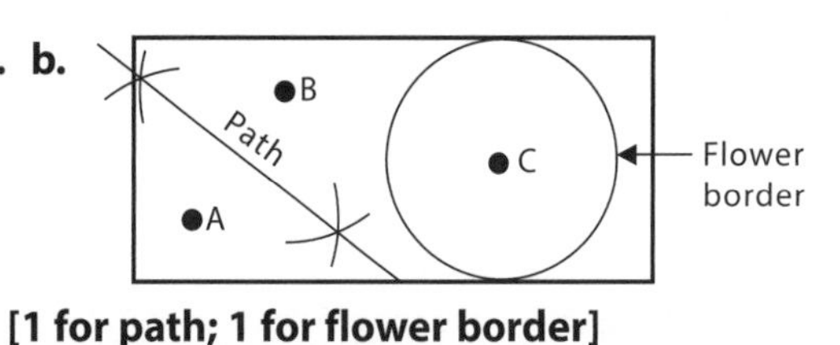

[1 for path; 1 for flower border]

Answers

pages 56–57

Angles

QUICK TEST

1. **a.** $a = 55°$

   b. $a = 74°, b = 32°$

   c. $a = 41°, b = 41°, c = 41°, d = 139°$

2. 30°

EXAM PRACTICE

1. For BE and CF to be parallel, the angles between the parallel lines are supplementary **[1]** and must add up to 180°.
   53° + 127° = 180° **[1]**

2. Total sum of interior angles = (6 – 2) × 180°
   = 720° **[1]**
   $y + 126° + 83° + 145° + 138° + 79° = 720°$ **[1]**
   $y + 571° = 720°$
   $y = 720° - 571°$ **[1]**
   $y = 149°$ **[1]**

3. Exterior angle of the hexagon = $\frac{360°}{6}$
   = 60°
   Interior angle of the hexagon = 180° – 60°
   = 120° **[1]**
   Exterior angle of the octagon = $\frac{360°}{8}$
   = 45°
   Interior angle of the octagon = 180° – 45°
   = 135° **[1]**
   $x + 135° + 120° = 360°$
   $x = 105°$ **[1]**

pages 58–59

Bearings

QUICK TEST

1. **a.** 072°

   b. 208°

   c. 290°

   d. 139°

EXAM PRACTICE

1. **a.** 098°

   b. 360° – 173° **[1]**

   = 187° **[1]**

2. $x = 180° - 143°$

   $x = 37°$ **[1]**

   Bearing = 360° – 37° **[1]**

   = 323° **[1]**

Drawing a sketch helps work out which angle is needed.

pages 60–61

Translations and Reflections

QUICK TEST

1. **a.** Reflection in $y = 0$ (x-axis)

   b. Reflection in $x = 0$ (y-axis)

   c. Translation of $\begin{pmatrix} -6 \\ 0 \end{pmatrix}$

   d. Translation of $\begin{pmatrix} 5 \\ -1 \end{pmatrix}$

EXAM PRACTICE

1. **a. b.**

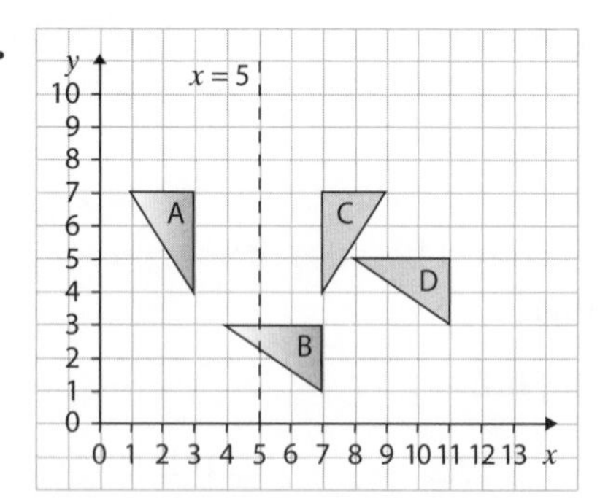

[2 for shape C; 1 for shape D]

   c. Reflection **[1]** in the line $y = x$ **[1]**

pages 62–63

Rotation, Enlargement and Congruency

QUICK TEST

1.

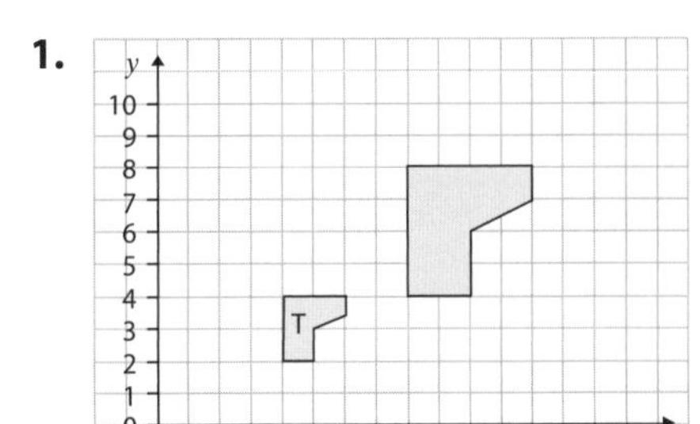

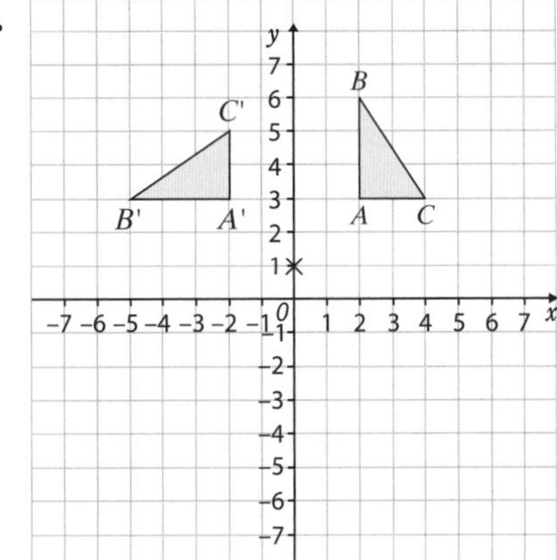

EXAM PRACTICE

1. There are several ways of proving this. An example might be:
Since triangle CDE is equilateral, length CD = length CE **[1]**. Both triangles have a common length, CF **[1]**, which is the perpendicular bisector of DE.
Hence length DF = length EF
By SSS, triangle CFD is congruent to triangle CFE. **[1]**

pages 64–65

Pythagoras' Theorem

QUICK TEST

1. a. 17.46 cm (2 d.p.)

b. 9.38 cm (2 d.p.)

EXAM PRACTICE

1. $26^2 = 24^2 + 10^2$ **[1]**

676 = 576 + 100 and this obeys Pythagoras' Theorem, so the triangle must be right-angled. **[1]**

2. Length of diagonal: $\sqrt{(5^2 + 3^2)} = 5.83\ldots$ **[1]**

Perimeter = 5 + 4 + 7 + 5.83…

Perimeter = 21.83… m **[1]**

Number of lengths: 21.83… ÷ 2.5 = 8.732

9 lengths of beading needed × £1.74 **[1]**

= £15.66 **[1]**

Always consider the context of the question. Here you round to 9 lengths of beading, since you would not be able to purchase 8.732 lengths.

3. $\sqrt{(17-7)^2 + (9-2)^2}$ **[1]**

$= \sqrt{149}$

= 12.2 **[1]**

Day 6

pages 66–67

Trigonometry

QUICK TEST

1. a. 5.79 cm

b. 8.40 cm

2. a. 38.7°

b. 43.0°

EXAM PRACTICE

1. a. $\tan PBN = \frac{\text{opp}}{\text{adj}}$

$\tan PBN = \frac{19.7}{12.6}$ **[1]**

$PBN = \tan^{-1}\left(\frac{19.7}{12.6}\right)$ **[1]**

Angle $PBN = 57.4°$ **[1]**

b. $\sin 48° = \frac{19.7}{PA}$ **[1]**

$PA = \frac{19.7}{\sin 48°}$ **[1]**

$PA = 26.5$ m (3 s.f.) **[1]**

pages 68–69

Trigonometry and Pythagoras Problems

QUICK TEST

1. 1.95 m^2 (3 s.f.)

Answers

EXAM PRACTICE

1. Let the length of the ladder be x.

   $x^2 = 3^2 + 0.6^2$ **[1]**

   $x^2 = 9 + 0.36$

   $x = \sqrt{9.36}$ **[1]**

   $x = 3.06\,\text{m}$ **[1]**

2. **a.** $\tan x = \frac{0.9}{4}$ **[1]**

   $x = \tan^{-1}\left(\frac{0.9}{4}\right)$ **[1]**

   $x = 12.7°$ **[1]**

   b. Length of sloping strut $= \sqrt{(0.9^2 + 4^2)}$ **[1]**

   $= \sqrt{(0.81 + 16)}$

   $= \sqrt{16.81}$ **[1]**

   $= 4.1\,\text{m}$ **[1]**

3. Let distance of town A from town C be x.

   $x^2 = 7.6^2 + 9.8^2$ **[1]**

   $x^2 = 153.8$

   $x = \sqrt{153.8}$ **[1]**

   $x = 12.4\,\text{km}$ **[1]**

   Angle ACB:

   $\tan C = \frac{7.6}{9.8}$ **[1]**

   $C = \tan^{-1}\left(\frac{7.6}{9.8}\right)$

   $C = 37.8°$ **[1]**

   Bearing $= 180° + 37.8°$

   Bearing of town A from town $C = 218°$ **[1]**

pages 70–71

Circles, Arcs and Sectors

QUICK TEST

1. **a.** Circumference = 31.42 cm (2 d.p.)

   Area = 78.55 cm^2 (2 d.p.)

   b. Circumference = 47.13 cm (2 d.p.)

   Area = 176.74 cm^2 (2 d.p.)

   c. Circumference = 43.99 cm (2 d.p.)

   Area = 153.96 cm^2 (2 d.p.)

2. Perimeter = 41.13 cm (2 d.p.)

   Area = 100.53 cm^2 (2 d.p.)

3. **a.** 14.05 cm

   b. 49.17 cm^2

EXAM PRACTICE

1. Area of rectangle: $110 \times 60 = 6600\,\text{m}^2$ **[1]**

   Area of circle: $\pi \times 30^2 = 2827.4\,\text{m}^2$

   Total area $= 9427.4\,\text{m}^2$ **[1]**

   Number of sacks of grass seeds $= \frac{9427.4}{275}$

   = 34.28… sacks of grass seeds **[1]**

   He needs 35 sacks of grass seeds. **[1]**

Again consider the context of the question. You need to round the answer up to find out the whole number of sacks of seeds needed.

pages 72–73

Surface Area and Volume 1

QUICK TEST

1. **a.** 38 cm^2

   b. 35.705 cm^2

2. **a.** 388.8 cm^3 (1 d.p.)

   b. 1292.3 cm^3 (1 d.p.)

EXAM PRACTICE

1. Area of the floor: $A = \frac{(a+b)}{2} \times h$

   $A = \frac{(4+6)}{2} \times 3$ **[1]**

   $A = 15\,\text{m}^2$ **[1]**

   Total cost $= 15 \times 155$ **[1]**

   = £2325 **[1]**

pages 74–75

Surface Area and Volume 2

QUICK TEST

1. **a.** 565 cm^3

   b. 637 cm^3

c. $905\,\text{cm}^3$

d. $245\,\text{cm}^3$

EXAM PRACTICE

1. $V = \pi \times r^2 h$

$205 = \pi \times 2.8^2 \times h$ **[1]**

$h = \frac{205}{\pi \times 2.8^2}$ **[1]**

$h = 8.32\,\text{cm}$ **[1]**

2. $\frac{1}{3}\pi r^2 h = 15$ **[1]**

$r^2 = \frac{15 \times 3}{\pi \times 2.1}$ **[1]**

$r = \sqrt{6.8209...}$ **[1]**

$r = 2.61\,\text{m}$ **[1]**

pages 76–77

Vectors

QUICK TEST

1. a. $\begin{pmatrix} 4 \\ 5 \end{pmatrix}$

b. (5, 4)

EXAM PRACTICE

1. a. i. $-\mathbf{a} + \mathbf{c}$

ii. $-\mathbf{c} - \mathbf{a}$

b. $\mathbf{a} + \frac{1}{2}(-\mathbf{a} + \mathbf{c})$ **[1]**

$= \frac{1}{2}\mathbf{a} + \frac{1}{2}\mathbf{c}$ **[1] or**

$\frac{1}{2}(\mathbf{a} + \mathbf{c})$

Day 7

pages 78–79

Probability

QUICK TEST

1. 0.35

2. 294 students

3. a. $\frac{4}{36} = \frac{1}{9}$

b. $\frac{27}{36} = \frac{3}{4}$

EXAM PRACTICE

1. Faulty components $= 0.03 \times 1200$ **[1]**

$= 36$ **[1]**

2. $0.04 + x + 4x + 0.23 + 0.19 + 0.28 = 1$ **[1]**

$5x = 1 - 0.74$ **[1]**

$5x = 0.26$

$x = 0.052$ **[1]**

pages 80–81

Tree Diagrams

QUICK TEST

1. a.

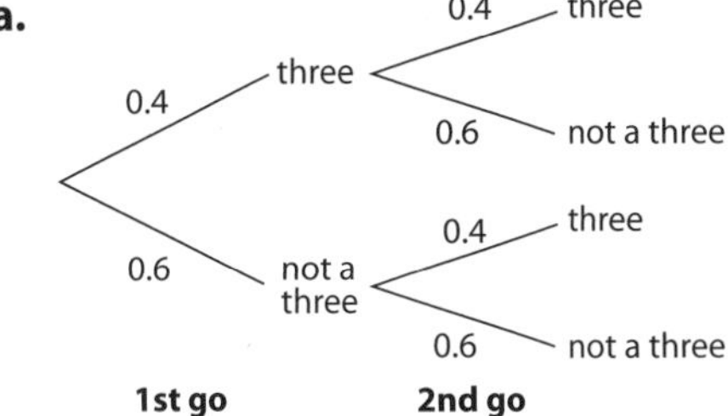

b. i. 0.16 ii. 0.48

EXAM PRACTICE

1.

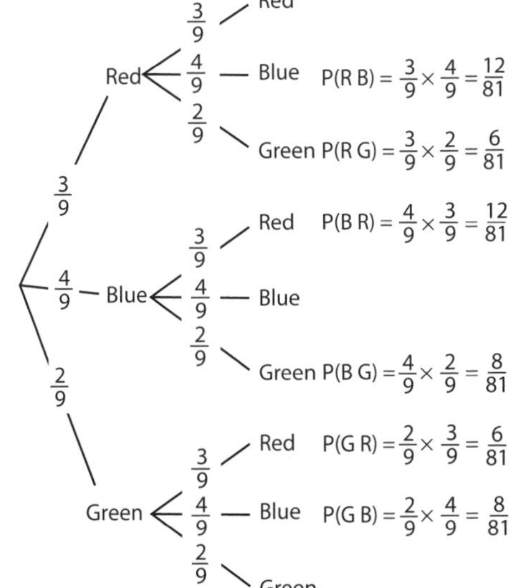

P(two different colours) $= \frac{12}{81} + \frac{6}{81} + \frac{12}{81} + \frac{8}{81} + \frac{6}{81} + \frac{8}{81}$ **[2]**

$= \frac{52}{81}$ **[1]**

Or 1 – P(same colours)

$= 1 - \left(\left(\frac{3}{9} \times \frac{3}{9}\right) + \left(\frac{4}{9} \times \frac{4}{9}\right) + \left(\frac{2}{9} \times \frac{2}{9}\right)\right)$ **[2]**

$= 1 - \frac{29}{81}$

$= \frac{52}{81}$ **[1]**

2. **Mr Smith** **Mrs Tate**

0.8 fiction — 0.4 fiction; 0.6 non-fiction

0.2 non-fiction — 0.4 fiction; 0.6 non-fiction

a. P(both take out fiction) $= 0.8 \times 0.4$ **[1]**
$= 0.32$ **[1]**

b. P(one of each type) $= (0.8 \times 0.6) + (0.2 \times 0.4)$ **[1]**
$= 0.48 + 0.08$ **[1]**
$= 0.56$ **[1]**

Although the question does not tell you to draw a tree diagram, doing so can help in working out the answers.

pages 82–83

Sets and Venn Diagrams

QUICK TEST

1. **a.** $C \cup D = \{2, 3, 4, 5, 6, 7, 8, 9\}$

   b. $C \cap D = \{6, 7\}$

EXAM PRACTICE

1. **a.**

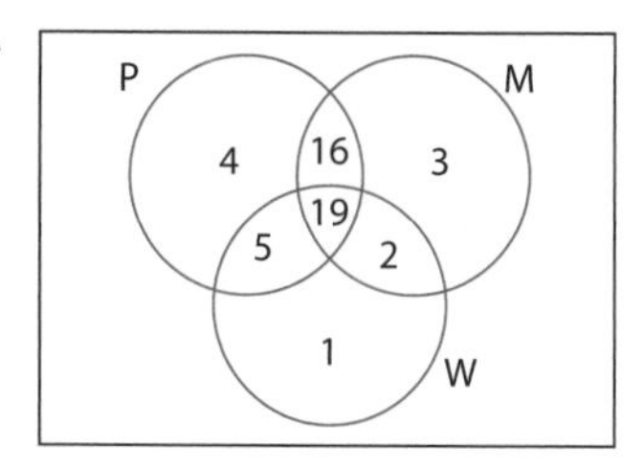

[1 for 19 in the overlap of all three ovals; 1 for four of the seven values correct; 1 for a fully correct Venn diagram]

b. $4 + 16 + 19 + 5 = 44$ **[1]**
P(plain chocolate) $= \frac{44}{50}$ **[1]** $= \frac{22}{25}$

c. $16 + 5 = 21$ **[1]**
P(likes one other type of chocolate) $= \frac{21}{44}$ **[1]**

pages 84–85

Statistical Diagrams

QUICK TEST

1. **a. i.** Roses

   ii. Carnations

   b. i. 5%

   ii. 15%

2.

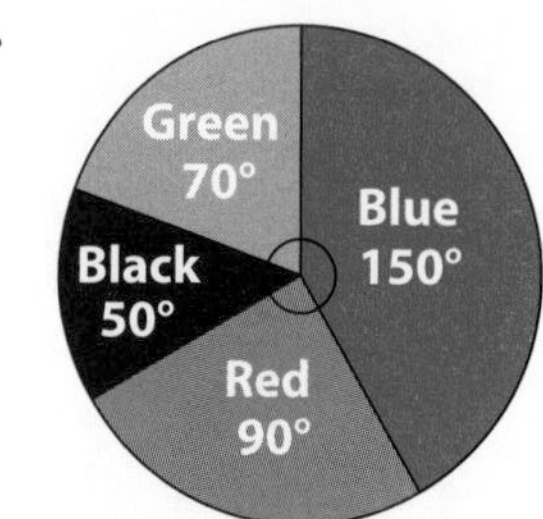

EXAM PRACTICE

1. Town D should have the retirement home since it has a much larger number of people aged 40 or over: 1100 people compared to 475 people in town C **[2]**. Town C should have the school since it has 200 0–19 year-old people compared to 150 people in town D aged 0–19 **[2]**.

pages 86–87

Scatter Diagrams and Time Series

QUICK TEST

1. **a.**

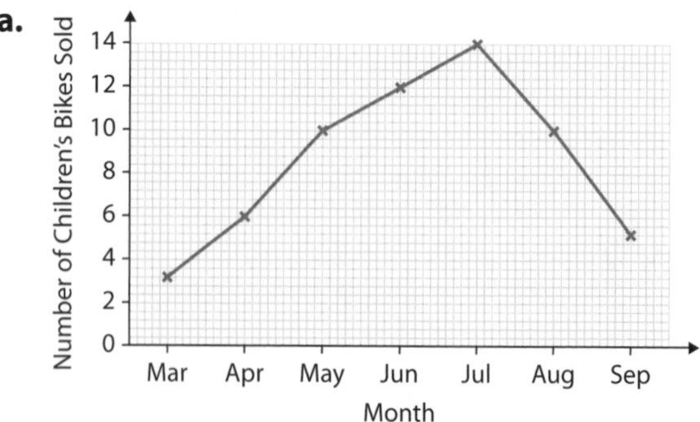

   b. The sales of the number of children's bikes increase between March and July and then the sales begin to decrease during August and September.

EXAM PRACTICE

1. **a.** Line of best fit should be as close as possible to all points and in the direction of the data.

   b. Approx. 2 years old
   [1 for evidence of attempt to read from line of best fit; 1 for correct answer]

   c. Approx. £3000
   [1 for evidence of attempt to read from line of best fit; 1 for correct answer]

   d. You cannot predict the value of a 10-year-old car because this is beyond the range of the data and the relationship may not hold.

pages 88–89

Averages 1

QUICK TEST

1. **a.** Mean = 4.75

   Range = 7

   b. Mean = 9.5

   Range = 18

2. **a.** Median = 2.5

   Mode = 2

   b. Median = 6

   Mode = 6

EXAM PRACTICE

1. **a.** Mean

   $= \frac{(4\times5)+(7\times6)+(10\times7)+(4\times8)+(3\times9)+(1\times10)}{(4+7+10+4+3+1)}$ **[1]**

   $= \frac{201}{29}$ **[1]**

   = 6.93 minutes **[1]**

   b. Median = 7

   c. Mode = 7

   d. Rupinder cannot be correct because the highest number of minutes taken to solve the problem was only 10 minutes, so the mean cannot be higher than this.

pages 90–91

Averages 2

QUICK TEST

1. 150.9

2. **a.**

0	7 9
1	8
2	5 5 5 6 7 7 8
3	1 1 1 3 6 6 7 9
4	0 2 2 7 9
5	0

   Key 4|2 = 42

   b. Median = 31

EXAM PRACTICE

1. Mean $= \frac{\Sigma fx}{\Sigma f}$

   Mean $= \frac{276}{50}$

   [1 for frequency × midpoint; 1 for totalling frequency; 1 for $\frac{276}{50}$]

   Mean = 5.52 hours **[1]**

Adding two extra columns to the table, i.e. midpoint (x) and fx, is helpful when working out the estimate of the mean.

Glossary

Alternate angles – angles formed when two or more lines are cut by a transversal. If the lines are parallel then alternate angles are equal.

Arc – a curve forming part of the circumference of a circle.

Arithmetic sequence – a sequence with a common first difference between consecutive terms.

Bearing – the direction measured clockwise from a fixed point. A bearing has three digits (for angles less than 100°, a zero, or zeros, is placed in front, e.g. 025°).

Bias – a tendency either towards or away from some value.

BIDMAS – an acronym that helps you remember the order of operations: Brackets, Indices and roots, Division and Multiplication, Addition and Subtraction.

Centre of enlargement – the point from which the enlargement happens.

Centre of rotation – the point around which a shape can rotate.

Chord – a line joining two points on the circumference of a circle.

Class interval – the width of a class or group, e.g. $0\,\text{g} < \text{mass of spider} \leqslant 10\,\text{g}$

Coefficient – a number or letter multiplying an algebraic term.

Common ratio – the ratio between two numbers in a geometric sequence.

Compound interest – interest that accrues from the initial deposit plus the interest added on at the end of each year.

Compound measure – a measurement using more than one quantity, often using 'per' as in speed, e.g. km/h

Congruent – exactly alike in shape and size.

Constant of proportionality – the constant value of the ratio of two proportional quantities x and y.

Correlation – the relationship between the numerical values of two variables, e.g. there is a positive correlation between the numbers of shorts sold as temperature increases; there is a negative correlation between the age and the value of cars.

Corresponding angles – angles formed when a transversal cuts across two or more lines. When the lines are parallel corresponding angles are equal.

Cross-section – the shape of a slice through a solid.

Direct proportion – two values or measurements may vary in direct proportion, i.e. if one increases, then so does the other.

Discrete data – data that can only have certain values in a given range, e.g. number of goals scored, shoe sizes.

Elevation – the 2D view of a 3D shape or object when looking at it from the side or front.

Empty set – a set containing no objects (members).

Enlargement – a transformation of a plane figure or solid object that increases or decreases the size of the figure or object by a scale factor but leaves it the same shape.

Equation – a number sentence where one side is equal to the other.

Expression – a statement that uses letters as well as numbers.

Exterior angle – an angle outside a polygon, formed when a side is extended.

Factorisation – finding one or more factors of a given number or algebraic expression.

Finite set – a set which has an exact number of members.

Formula – an equation that enables you to convert or find a value using other known values, e.g. area = length × width

Geometric sequence – a sequence with a common ratio.

Gradient – the measure of the steepness of a slope:

$$\frac{\text{vertical distance (change in } y)}{\text{horizontal distance (change in } x)}$$

Highest common factor (HCF) – the highest factor shared by two or more numbers.

Identity – an identity is similar to an equation, but is true for all values of the variable(s); the identity symbol is $\equiv$

e.g. $2(x + 3) \equiv 2x + 6$

Imperial (units) – traditional/old units of weight and measurements, which have generally been replaced with metric units.

Independent events – two events are independent if the outcome of one event is not affected by the outcome of the other event, e.g. tossing a coin and throwing a dice.

Index (also known as power or exponent) – the small digit to the top right of a number that tells you the number of times a number is multiplied by itself, e.g. 5^4 is $5 \times 5 \times 5 \times 5$; the index is 4.

Inequality – a statement showing two quantities that are not equal.

Infinite set – a set which goes on and on, i.e. it has no end number.

Integer – any whole number, positive or negative, including zero.

Intercept – the point where a line or graph crosses an axis.

Interior angle – an angle between the sides inside a polygon.

Intersection – the point at which two or more lines cross.

Inverse (indirect) proportion – two quantities vary in inverse proportion when, as one quantity increases, the other decreases.

Like terms – algebraic terms that are the same, apart from their numerical coefficients, e.g. $2d$ and $6d$.

Linear sequence – a number pattern which increases (or decreases) by the same amount each time.

Line of best fit – a line (usually straight) drawn through the points of a scatter diagram, showing the trend, which enables you to estimate new values using original information.

Line of symmetry – a line that splits a 2D shape into two equal halves.

Locus (plural: loci) – the locus of a point is the path taken by the point following a rule or rules.

Lower bound – the bottom limit of a rounded number.

Lowest (least) common multiple (LCM) – the lowest number that is a multiple of two or more numbers.

Mean – an average value found by dividing the sum of a set of values by the number of values.

Median – the middle item in an ordered sequence of items.

Metric (units) – units of weight and measure based on a number system in multiples of 10.

Midpoint – the point that divides a line into two equal parts.

Modal class – the largest class in a grouped frequency table.

Mode – the most frequently occurring value in a data set.

Multiplier – the number by which another number is multiplied.

Mutually exclusive events – two or more events that cannot happen at the same time, e.g. throwing a head and throwing a tail with the same toss of a coin are mutually exclusive events.

Net – a surface that can be folded into a solid.

Parallel – lines that stay the same distance apart and never meet.

Percentage increase / decrease – the change in the proportion or rate per 100 parts.

Perpendicular bisector – a line drawn at right angles to the midpoint of a line.

Plan view – the 2D view of a 3D shape or object when looking down onto it.

Population – any large group of items being investigated.

Power – the small digit to the top right of a number that tells you the number of times a number is multiplied by itself, e.g. 5^4 is $5 \times 5 \times 5 \times 5$.

Prime factor – a factor that is also a prime number.

Prime number – a number with only two factors, itself and 1.

Prism – a 3D shape that has a uniform cross-section.

Probability – the probability of an event occurring is the chance that it may happen, which can be expressed as a fraction, decimal or percentage.

Glossary

Pythagoras' Theorem – the theorem which states that the square on the hypotenuse of a right-angled triangle is equal to the sum of the squares on the other two sides.

Quadratic equation – an equation containing unknowns with maximum power 2, e.g. $y = 2x^2 - 4x + 3$. Quadratic equations can have 0, 1 or 2 solutions.

Quadratic graph – the ∪ shaped graph of a quadratic equation.

Random sample – a sampling method in which each data object/person has an equal chance of being selected.

Range – the spread of data; a single value equal to the difference between the greatest and the least values.

Ratio – the ratio of A to B shows the relative amounts of two or more things and is written without units in its simplest form or in unitary form, e.g. $A : B$ is 5 : 3 or $A : B$ is 1 : 0.6

Reciprocal – the reciprocal of any number is 1 divided by the number (the effect of finding the reciprocal of a fraction is to turn it upside down), e.g. the reciprocal of $\frac{2}{3}$ is $\frac{3}{2}$

Recurring decimal – a decimal that has digits in a repeating pattern, e.g. 0.3333 or 0.252 525.

Reflection – a transformation of a shape to give a mirror image of the original.

Relative frequency – $\frac{\text{frequency of a particular outcome}}{\text{total number of trials}}$

Resultant – the result of adding two or more vectors together.

Roots – in a quadratic equation $ax^2 + bx + c = 0$, the roots are the solutions to the equation.

Rotation – a geometrical transformation in which every point on a figure is turned through the same angle about a given point.

Sample – a section of a population or a group of observations.

Sample space diagram – a probability diagram that contains all possible outcomes of an experiment.

Scalar – a quantity which has only magnitude.

Scale factor – the ratio by which a length or other measurement is increased or decreased.

Scalene – a triangle that has no equal sides or angles.

Scatter diagram – a statistical graph that compares two variables by plotting one value against the other.

Sector – a section of a circle between two radii and an arc.

Set – a collection of objects (members).

Similar – the same shape but a different size.

Simple interest – interest that accrues only from the initial deposit at the start of each year.

Simplify – making something easier to understand, e.g. simplifying an algebraic expression by collecting like terms.

Simultaneous equations – two or more equations that are true at the same time. On a graph the intersection of two lines or curves.

Standard index form (Standard form) – a shorthand way of writing very small or very large numbers; these are given in the form $a \times 10^n$, where a is a number between 1 and 10.

Stem and leaf diagram – a diagram used for displaying data by splitting the values.

Stratified sampling – a sampling method where the population is divided into categories and a sample is taken using the same proportion in each category as in the whole population.

Subset – a set within a set.

Substitution – to exchange or replace, e.g. in a formula.

Supplementary angles – angles that add up to 180°.

Surface area – the area of the surface of a 3D shape, equal to the area of the net of that shape.

Tangent – a straight line that touches a curve or the circumference of a circle at one point only.

Term – in an expression, any of the quantities connected to each other by an addition or subtraction sign; in a sequence, one of the numbers in the sequence.

Terminating decimal – a decimal fraction with a finite number of digits, e.g. 0.75

Theoretical probability – a predicted probability.

Translation – a transformation in which all points of a plane figure are moved by the same amount and in the same direction. The movement can be described by a vector.

Tree diagram – a way of illustrating probabilities in diagram form. It has branches to show each event.

Trigonometry – the branch of mathematics that shows how to explain and calculate the relationships between the sides and angles of triangles.

Turning point – in a quadratic curve, a turning point is the point where the curve has zero gradient. It could be a minimum or a maximum point.

Universal set – contains all the objects being discussed.

Upper bound – the top limit of a rounded number.

Vector – a movement on the Cartesian plane described using a column, e.g. $\begin{pmatrix}3\\4\end{pmatrix}$

Venn diagram – a diagram used to represent sets.

Vertex – in 2D, a point where two or more lines meet; in 3D, the corners of a shape, where the edges meet.

Notes

Notes

ACKNOWLEDGEMENTS

The author and publisher are grateful to the copyright holders for permission to use quoted materials and images.

Cover & P1: © Andrus Ciprian/Shutterstock.com

All other images are © iStock/Thinkstock/Getty Images; © Shutterstock.com and © HarperCollins*Publishers* Ltd

Every effort has been made to trace copyright holders and obtain their permission for the use of copyright material. The author and publisher will gladly receive information enabling them to rectify any error or omission in subsequent editions. All facts are correct at time of going to press.

Published by Letts Educational
An imprint of HarperCollins*Publishers*
1 London Bridge Street
London SE1 9GF

ISBN: 9780008165949

First published 2016

10 9 8 7 6 5 4 3 2 1

British Library Cataloguing in Publication Data.
A CIP record of this book is available from the British Library.

Commissioning Editor: Emily Linnett
Author: Fiona Mapp
Project Management: Richard Toms
Cover Design: Paul Oates
Inside Concept Design: Ian Wrigley
Text Design and Layout: Q2A Media
Production: Lyndsey Rogers
Printed in China